Praise for *Peat and Whisky*

"This is an outstanding co[...] indeed I believe it to be among the most impor[...] whisky ever written." —**Charles MacLean**, noted whisky writer and connoisseur

"Compelling and entertaining … a vast work of research, and clearly a labour of great love. It should be essential reading for all whisky lovers." —**Gavin D Smith**, whisky writer

"Fascinating … thorough, engaging and charmingly written … *Peat and Whisky* is not only a joy to read, it also makes a vital contribution to understanding our precious environment." *The Keeper* (Keepers of the Quaich magazine), Winter 2023

"If you consider yourself a connoisseur of the amber nectar, this is a compulsory read." *The Ìleach*

"A unique work … essential reading for anyone with an interest in whisky." —**Neil Wilson**, whisky historian, writer

"Peat has had a vital, if unappreciated, place in Scottish life and the whisky industry for centuries … A timely book, indeed one long overdue." —**Brian Townsend**, whisky historian and writer

"Groundbreaking … In this essential read, Billett fuses landscape, history, culture, and a dash of alchemy to unpack how peat fuels Scotland's liquid gold." **WineDharma.com**

"A much-needed dissection of the history of one of Scotch whisky's most important ingredients … A must-read for any whisky fan, especially as we enter a time where the use of peat to create 'just' a drink is being more closely examined." —**Billy Abbott,** whisky ambassador at The Whisky Exchange

"An inspirational work, wonderfully engaging, educational and thought-provoking … answers all the questions you could possibly ask about peat, while tying this all into a compelling read for any Scotch whisky enthusiast." —**Iain J McAlister**, Master Distiller, Glen Scotia, Campbeltown

PEAT AND WHISKY
THE UNBREAKABLE BOND

MIKE BILLETT

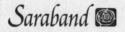

Saraband

Published by Saraband
3 Clairmont Gardens
Glasgow, G3 7LW
www.saraband.net

ISBN: 9781913393908

2 3 4 5 6 7 8 9 10

Printed and bound in Great Britain by Clays Ltd, Elcograf S.p.A.

Contents

To Sofia

Orkney
Islands
Eday

Thurso

Kylesku
Flow
Country
Wick

Isle of
Lewis

Brora

Inverness
Ben Rinnes

Skye
R. Spey
Aberdeen

Ben Nevis

Glasgow
Edinburgh

Islay

Campbeltown

Glossary

abv   Alcohol by volume, expressed as a percentage

afforestation   Planting of trees on land that was not formerly used for forestry

anaerobic   Free of oxygen

anthracite   Highest grade, most carbon- and energy-rich form of coal

blackhouse   Traditional, stone or turf-built cottage found in Ireland or the Scottish Highlands and Islands. Often built without a chimney

blanket bog   Large expanse of peat formed in areas of high rainfall and low evaporation/transpiration

blanket forest High density planting of trees, typically conifers, that creates a dark understory with no light penetration, or ground vegetation

bourbon   Sweet style of American whisky distilled from a mixture of grains that contain at least 51% corn

BP   Before present. The end of the ice age and start of the Holocene period is dated at 11,700BP

case (of whisky) Box or crate usually containing 12 bottles

charcoal   Black carbon-rich material formed by heating wood or peat in an oxygen-starved environment to remove water and volatiles

char   Surface layer inside a barrel created by firing to produce a range of flavours in the maturing spirit

chromatography Analytical method used to separate individual components of a mixture by passing them in a mobile (gas or liquid) phase through a stationary phase

chlorophyll   Green plant pigment that converts light to chemical energy by a process known as photosynthesis

coke   Carbon-rich fuel made by heating coal in the absence of air to drive off impurities

PEAT AND WHISKY

condenser	Equipment that converts the spirit vapour produced by distillation to a liquid by cooling
congener	Compound produced during whisky making that creates aroma and flavour
cut point	Specific alcohol content when a distiller switches between foreshots, spirit or feints during the spirit run
deforestation	Permanent removal of trees from land
distillation	Heating process that separates individual components of a mixture by exploiting differences in their volatility
draff	Remains of malted barley at the end of mashing that is often dried and used for animal feed
dross	Fine dusty or fibrous fragments of dried peat. Sometimes called 'caff'
dubh lochan	Gaelic for 'small black loch', often associated with peatlands
fermentation	Chemical process by which sugars are broken down by enzymes in the absence of oxygen. In whisky production this involves the use of yeast to make alcohol
frack(ed) gas	Gas produced by man-made hydraulic fracturing of bedrock. Often called 'shale gas'
furfural	Liquid aldehyde compound produced during whisky making with an almond-like, sweet, biscuity aroma and flavour
gauger	Colloquial term used in Scotland and Ireland for an exciseman or customs agent
grain whisky	Whisky made from any type of grain in a column (or Coffey) still by a process of continuous distillation
greenhouse effect	Rise in temperature of the lower atmosphere caused by 'greenhouse gases' such as carbon dioxide and methane
greenwashing	Creating the impression that a company is more environmentally friendly than it actually is

GLOSSARY

grist	Rough, ground malt produced prior to the mashing stage
hogshead	Cask of approximately 250 litres usually made from American oak
Holocene	Period of geological time that started at the end of the last ice age
humification	Decomposition of soil organic matter
hydrophobic	Literally 'water hating', in contrast to 'hydrophilic' (water loving)
isotope	Form of the same element that contains a different number of neutrons
lignin	Large, complex polyphenolic molecule that makes plant cell walls strong and 'woody'
Macfarlane Method	Colorimetric analytical technique used to measure total phenol content of malted barley. Colorimetric methods are based on a relationship between absorption of light and concentration
malted barley	Often simply called 'malt', barley that after being soaked in water has had its germination halted by drying. This is done by 'maltmen' or 'maltsters' in a 'maltings' or 'malthouse'
malt whisky	Whisky made only from malted barley
mashing	Process of mixing the grist with hot water to extract the fermentable sugars. Mashing takes place in a 'mash tun'
mass spectrometry	Analytical technique that separates and identifies molecules based on differences in their charge and mass
maturation	Aging of spirit in casks to create a range of aromas and flavours that contribute significantly to the character of the final whisky
metamorphic rock	Formed by the transformation of existing rocks by heat, pressure or both. Slate, schist and gneiss are all examples of metamorphic rocks

minerotrophic	Bog type formed by, and associated with, mineral-rich waters. Typically, nutrient-rich and species-rich
new-make spirit	Product of the spirit still - colourless and of high alcoholic strength
olfactory	Relating to the sense of smell
ombrotrophic	Acidic, nutrient-poor type of bog that derives its water largely from precipitation
pedagogical	Related to teaching or methods of education
peatland degradation	Damage and destruction of peatland habitat, hydrology and function. Degraded peatlands can be regenerated by restoration
peat hag	Overhanging, eroded peat bank or gully
perch	Traditional measurement used in peat cutting. A perch was a five-metre square area of hand-cut peat turfs laid out on the surface
phenolics	Molecular compounds with one or more aromatic rings of six carbon atoms. They include large polyphenolic molecules like lignin, or smaller monophenols made up of one single ring. Guaiacol, phenol and cresol and all examples of monophenolic compounds
ppm	Parts per million. Unit of measurement commonly used to determine the concentration of phenols in malted barley and less frequently in the distilled spirit
pot still	Rounded, pot-bellied vessel used to make malt whisky and other spirits by batch distillation
pyrolysis	Combustion in an oxygen-starved environment with low airflow and relatively low temperature
refill	Term used to describe the second filling of a cask that previously held whisky
sedge	Grass-like plant with a triangular stem and leaves arranged in spirals
serendipity	Accidental, unplanned, but fortuitous discovery
sheepfold	Pen or enclosure for sheep

GLOSSARY

shieling	Rough, usually stone-built, open shelter for protecting animals in remote places
sphagnum	Group of almost 400 individual species of moss. A key peat-forming plant group; often termed the 'Super Mosses'
splint coal	Hard, dull coal that fractures and splits along planes of weakness
sma' stell	Scots name for the small stills that were once used to make illicit whisky
still	Heated vessel used to distil liquids typically made of copper. In malt whisky distilleries it operates in pairs of wash still (first distillation) and spirit still (second distillation)
tephra	Rock fragments or particles ejected into the atmosphere by a volcanic eruption
tor	Free-standing, conspicuous rock outcrop that typically forms by differential weathering on the summits of hills or ridges
treisgear	Traditional Gaelic name for a peat cutting tool. Sometimes called *tarasgeir* or *tairsgeir*
vatted malt	Marriage of casks of single malt whiskies produced from different distilleries
volatilisation	Process in which a liquid becomes a vapour
washback	Large wooden or steel container where sugary wort is fermented to alcohol by the addition of yeast
wash	Fermented alcoholic liquid produced in the washback and collected in the wash charger prior to distillation
watershed	Land area with specific boundaries that defines the source region of individual rivers or lochs. Also known as a 'catchment'
wormtub	Old fashioned type of condenser with a coiled copper tube inside
wort	Liquid produced by mashing and often likened to sweet tea

Introduction

by Dave Broom

How many people have smelled a peat fire? How many have cut peat, even seen it? It was alien to me, growing up in Glasgow. It was the smell of another Scotland, a scent of places and a culture barely known. How significant peat was in those places can be gleaned from language. The Scots Thesaurus has more terms for rain than it does for sun. This is replicated with peat. Finlay MacLeod's *A Peat Glossary* lists 120 terms for peat and peat bogs gathered on Lewis. The depth and precision of the terms reflect the links with man and land: naming, noticing. Peat was part of this world. Its reek was in the drink and your clothes and your world.

As I began to write about whisky, so I began to learn how to cut peat (badly). My tutors were old whisky men: Norrie Kimble and Iain McArthur on Islay, Norman MacLeod on Skye, a whole team on Orkney. There were differences in terminology, ways of cutting, even tools, but all of them spoke of being on the moss in the summer, cutting for the next distilling season: the hard work, the midges, the bottles smuggled out of the distillery stashed in the peat bank. Cutting peat was part of the old rhythm of whisky making, the heartbeat of distilling.

'I want it to have the smell of peat in the village on a soft evening,' one Hebridean distiller told me, describing the intended aroma of his new whisky. A thread of blue smoke linking the spirit with the place of its birth.

Peat has warmed and dried, coloured burns, lochs (and baths). It's been a medicine, has aromatised whisky and preserved food, yet this claggy, damp, black and brown matter that drapes itself over Scotland remains mysterious. Until now.

PEAT AND WHISKY

Mike Billett has spent over twenty years as a peatland scientist. He is the perfect person to guide us into peat's fascinating story. Distilling is about concentrating flavour, so too is what is contained and compressed within a peat bank. Like a spade cutting into the ooze, he reveals the complexity of what lies beneath, the deep layers of stories awaiting to be revealed.

It is easy to become bogged down (literally in this case) when writing about science, but these accounts are never less than easy to understand and engaging. A self-confessed geophagist (eater of earth), he gives us tasting notes for different peats (as well as the ash from the old Brora kiln).

He drinks peaty water, tastes smoky whiskies, reveals the complexities of this world and how it has been a vital element not just for whisky, but Scotland itself. In his hands, peat is about more than phenols, but a memory bank built up over millennia, its stories finally released.

In his telling of peat's story, he gives us an alternative, and necessary, story of whisky. The use of peat may have started as expediency – how to dry barley and heat stills when there were no trees or coal, but by the seventeenth century it had become, in the words of Sir Robert Moray, 'the best fewell for malting' because of its quality and aromatic properties. It had become part of the recipe, part of whisky's identity. Even today, Scottish peated barley goes around the world.

By the nineteenth century, as Mike outlines, Scotland was a carbon-driven economy whose success and prestige was built on exploitation of resources. Places such as Eday in Orkney had an economy entirely based on peat.

In his telling, peat can be used to measure the changing fortunes (and flavour) of whisky. He shows how, as the industry changed and grew larger, so local mosses were abandoned, and then through the drive for efficiency and a shift in palate preferences, the smokiness of many whiskies began to drop. The irony is that it is the notable exceptions to this, particularly the smoky whiskies from Islay, which have disproportionately fuelled the boom in single malt. Today, many people's first whisky will be peated.

Carried within this peaty wave has come a variety of misconceptions that are gently corrected here, such as peaty water doesn't make a whisky smoky. He also calls out, politely, the 'rather odd convention' of using measurements of phenolic parts per million (smokiness) in the barley rather than the final product, a practice that is absurd given that 60–80% of phenols can be lost during the whisky-making process. Will this setting right result in whisky marketeers finally stopping this practice? I hope so.

The role of people – for good and otherwise – is central to the telling of this story, and to do that he embarks on a series of evocatively described walks. You are beside him as he ventures into bogs, along the lonely and unused peat roads, urban edgelands and post-industrial wastelands, revealing forgotten stories of places long lost: Slamannan and Fannyside, Eday, New Pitsligo, Faemussach and Birnie.

This book then, appropriately enough, is about exhumation. Peat preserves. Now is the time for it to reveal what it holds. He suffers for his art – walking through storms, drenched to the skin, clambering through ditches, watching Brora Rangers play while standing in a former peat shed.

The heart of the story is an epic trek across the Flow Country, east to west, coast to coast, an apparently blank space that he gives new relevance and importance as its geology, landscape, and history is shown, and its vital role in peat's future is revealed.

It is also a story about exploitation. You shudder on reading his description of the 'far-famed' Faemussach moss in Speyside and how its ten-metre-deep beds were scraped clean. Part of the reason of using peat as fuel in the first place was only because Scotland had been deforested. As a species, we always push things too far. Peat, went the thinking, is an inexhaustible resource. Find it, scalp it, or drain it, degrade it, then move on.

That attitude is now, thankfully, changing. This is also a book about preservation and reclamation. He shows us not a featureless landscape but a living one, rich with flora, alive with birdsong, one which breathes.

It is this inhalation that is central to peat bogs' importance. Peat, in Mike's words, 'is nature's own carbon capture and storage system'. It locks in carbon more efficiently than trees, and retains water, 'buffering the hydrological system'. If degraded and dried out, however, not only is all of the carbon released, but the risk of flooding is increased. He tells us that 60 per cent of Scotland's distilleries are located in catchments containing important peat mosses, which supply and control water flow.

Recently, I was criticised by a distiller for suggesting how important peat restoration was as 'there are more urgent priorities in terms of reducing carbon emissions. After all, we only take a tiny amount of peat.' In Mike's estimation, the whisky industry removes less than 3 per cent of the peat extracted annually, but after the proposed ban on extraction for garden compost is enforced in 2024, the spotlight will suddenly be on the whisky industry. Whatever the case, peat restoration is important not just for whisky, but the planet.

He is optimistic. As he writes, 'a subject few people cared about or even noticed' is now beginning to be understood. He sees a 'seismic shift' in perception and action, and highlights how it is Scotland that is leading the way in peat science and restoration techniques, something which the whisky industry should draw on – and help.

Even as he walks through the Faemussach or the edgelands, he sees life returning, peat being laid down, the landscape coming alive once more.

INTRODUCTION

'I have spent twenty years researching the peatlands in this part of the world, now it was time to journey; to walk, look, see, listen, learn, and escape,' he writes. I would add – and also teach and guide, so we can learn. This is a vitally important and beautifully written book. Draw close to the fire, pour a dram, and listen to the breathing of the land.

Dave Broom, Hove
June 2023

One

The Story of a Piece of Peat

'The secret of the Islay whisky is the peaty waters and its peat.'
Bessie Williamson, Laphroaig Distillery Manager, 1954–72[1]

Port Ellen, Islay

The two of us are walking in bright early morning sunshine on the sands of Kilnaughton Bay, a mile west of Port Ellen. Paddy smells it moments before I do. He stops whatever he was doing, puts his nose to the air, flares his shiny black nostrils and promptly sneezes. A rapid shake of the head removes the offending odour from his airways and he quickly moves on. His sense of smell is up to 100,000 times more powerful than mine, and soon I too detect the soft aroma of burning, brought low across the bay in an invisible, meandering plume. I pause while strong sensory signals start flashing away inside my brain, and then the synapses fire, connecting past and present. Grilled breakfast kippers, the dying embers of last night's beach bonfire, a morning bacon roll, newly laid tarmacadam? A pub in Dingle, County Kerry? Maybe, but this is unmistakeably the smell of Islay, the 'Green Isle' on Scotland's west coast – the spiritual home of peated whisky. Measuring twenty-six miles long and eight miles wide and shaped like some wonky piece of jigsaw puzzle, it also goes by the name of the Peat Isle: almost a quarter of its land surface is wrapped in a blanket of wet, precious peat. It is currently home to nine working distilleries, soon to be twelve and maybe more.

An hour later, the kiln door swings open to reveal the largest bonfire of peats we have ever seen, a plume of intense light grey smoke billowing upwards from its surface. A fresh draft of air is

7

sucked through the open door, oxygenating the fuel, and the fire suddenly flares and bursts into life. Not good. Unskilled at kiln work and with differing levels of success, we take turns heaving shovel loads of dried peat into Kiln No 3. Each one of us backs off quickly to watch sparks fly and feel the peat heat. Smoke, not heat, is required to add flavour to the malt, but this is lost on us for the present. Our faces are illuminated by the light of the fire — peat-filled spade in hand, we pose for photographs in the glow. The kilnsman throws a small cube of yellow elemental sulphur onto the fire. Mined from some active volcano far from these shores, it quickly combusts and iridescent blue drops of molten sulphur drip like lava onto the white ash below.

We move outside the kiln room and, dazzled by the bright morning sunshine, look up at the dense plume of peat smoke rising energetically into the clear blue sky. Next stop on our itinerary is a big well-ventilated peat shed, largely empty except for a bank of newly arrived dark brown turfs hiding in the shade at the back. If the weather holds, the shed will soon be full of this year's harvest, which is currently drying in the warm sea air flowing over the surface of Castlehill Moss. Their story will end soon with a short journey across the flat peatlands of Islay to the maltings at Port Ellen.

All week, the smell and view of the maltings were a constant in our home on The Ard peninsula, overlooking the harbour and ferry terminal. When I woke in the morning and looked out through my bedroom window, a plume of white smoke would already be rising from one of the three chimneys into the calm air above the town. A spell of high-pressure weather over Scotland's Atlantic coast had squeezed and stabilised the lower atmosphere, making air movement difficult. When a late morning or afternoon breeze set the plume of peat reek in motion, it wandered gently around the bay, intact and close to the tops of trees and roofs. At the end of the day, when the heat subsided and the wind dropped, the plume would disentangle itself from the land and return to vertical, its dense column of smoke particles refracting the evening sunlight and turning the sky an even deeper red. We enjoyed a bottle of peated Finlaggan single malt one night, as the sun dropped below the horizon and the lights and midges of Port Ellen came out. Paddy looked on and gave us that canine stink eye look.

Fèis Ìle 2018 was blessed with day after day of not just warm but hot sunny weather. Whatever the weather, and every year since 2000, the communities of Islay and close neighbour Jura come together with the help of music and food to celebrate their whisky and distilleries with friends and relatives. This week in May is also a time when Ìleachs (natives of Islay) warmly welcome thousands of whisky pilgrims from all corners the world to their shores, and the island's 3,000-strong population is swollen by up to 15,000 new guests. Back in the 1960s, tourism at this scale was something new to Islay. *Whisky Island*, a black-and-white documentary film[1] shot for Scottish Television more than fifty years ago, shows there was much debate amongst the islanders about the benefits and dangers of tourism. At the time, the annual tourist season only lasted six weeks, but a glowing article in a Sunday newspaper in 1964 led to a surge in tourism that briefly stretched the island's resources and facilities. That summer, there wasn't a bed to be had and local people were worrying about the arrival of holiday camps, and their island home becoming the next Jersey, Isle of Man or, God forbid, Blackpool.

Even though it is hard to find a spare bed during *Fèis Ìle* week and busy roads and ferries are still hot topics of conversation, Islay accommodates and warmly receives visitors all year round, many attracted by its large and growing number of distilleries. Even in winter, the island has a special atmosphere and visitors quickly forget their urban mainland inhibitions. The people of Islay are open, friendly and welcoming. There are often a few extra words, maybe a smile and even the Islay – hands on the steering wheel, please – vehicle wave, or the less energetic version – a nonchalantly raised index finger. During *Fèis Ìle* week, the islanders have sometimes to be even more accommodating. Georgie Crawford, once of Lagavulin and now manager of Islay's new Portintruan Distillery, described how 'One morning, I came out of my front door to find someone camping in my front garden and cooking sausages for breakfast. They looked up, said "Hi, Georgie" and waved. I had no idea who that person was.'[2]

For the *Fèis Ìle*, most distilleries put on a new coat of paint, sweep the yard clean and in turn throw open their doors to crowds of whisky fans, and particularly fans of peated whisky, the Islay house style. Listen to the words of Bessie Williamson, UK Woman of the Year in 1953 and manager of Laphroaig Distillery, speaking in the mid-1960s on the *Whisky Island* documentary film: 'The secret of the Islay whisky is the peaty waters and its peat.' Well-spoken, popular, charming and gazing straight at the reporter through winged, cat-eye spectacles, looking more Audrey Hepburn than distillery manager, she goes on to sound a word of caution: 'Islay whisky, by itself, is rather too powerful for most people.'

The week was so hot that tarmac on the new cycle path connecting Port Ellen to the three distilleries on the Kildalton coast started to flow at its edges and give off that rather appealing and addictive smell of fresh bitumen. Groups of whisky fans who had set out earlier in the day at a brisk pace in the cool of the morning to either Lagavulin, Laphroaig or Ardbeg would return in the late afternoon sun at a more leisurely pace. Overcome by the island's whisky, some succumbed to the temptation of a short afternoon snooze under a tree or alighted in the shade of a strategically positioned park bench. A group of young mums and their kids sold iced water and soft drinks for charity to tiring foot soldiers on their long march back to barracks.

The firework display on the opening night in Port Ellen had set the dry grass ablaze on The Ard and called out the local fire engine from Bowmore to deal with this unplanned combustion. The beaches at Islay's coastal distilleries hosted family picnics, pebble skimming competitions, swimming dogs, paddling, and group dramming. Dress code was strictly dark glasses, sun hats, shorts and short sleeves. Islay simply looked and felt glorious. Long days, pink flowering sea thrift covering the cliffs, big sunsets, happy people, late nights with the occasional song on the beach, and the thought of the same to follow the next day, and the day after.

Sofia and I swam in Port Ellen harbour and off Tràigh Bhàn, the Singing Sands. Porters Family Butcher in Bowmore was doing a roaring trade in venison burgers and pork link sausages; the Co-op in Port Ellen ran out of barbecue charcoal. Midges harassed a lively, happy queue outside the Nippy Chippy mobile fish and chip shop on its regular Friday evening call by the beach at Port Ellen. Sailing ships arrived to use the distillery jetties, originally built for the coastal puffers that once brought grain and coal to the island before returning to the mainland with casks full of Islay whisky. We tasted whisky on a ship, in warehouses, on shingle and sandy beaches and even in a tractor shed, but never under cover in a bar.

No rain fell during the *Fèis Ìle* 2018. Out on the mosses, the bare peat became cracked, dry and dusty; stream levels dropped further, and Islay's famous brown peaty water became more intense in colour. It was perfect harvesting weather.

Castlehill Moss, Islay

I did not visit Castlehill Moss in that hot, dry May, but some years later on a new road built across peatland, I finally got to set eyes on the peat bog that is the main source of the taste of Islay whisky. From this single bog, 2–3,000 tonnes of dry peat are harvested each year and trucked four short miles to Diageo's Port Ellen Maltings.

PEAT AND WHISKY

Only the most hardened whisky tourist or peat geek ever ventures into the outback of Islay's remote peatlands, but up before breakfast, high above the shores of Loch Indaal, the two of us walk inland along a lonely road. At its end lies a community wind turbine and beyond a track disappears into the void of Islay's interior towards its remote fishing lochs and stalking grounds. Paddy's on the long lead and as we climb gently, almost imperceptibly uphill, he raises his head and periodically stops, concentrating on the scent of deer drifting across the open moorland. We pass the occasional solitary tree; the skylarks are up, crows do what crows do at this early hour and cuckoos call out to attract prospective mates. As we walk further inland, the peat thickens in the roadside cuts. At its base is a pale boulder clay containing assorted bright white fragments of quartzite, some as large as people, others no more than gravel. Most are what geologists call sub-rounded, telling me they have been moved, not far, by ice and water from the island's interior towards the sea. At the height of the last Ice Age, Islay was buried under half a kilometre of ice[3]. By 15,000BP,[4] the ice had gone and vast glacial deposits of rock, gravel and mud covered the island. As the climate warmed, the tundra melted, forests became established and hunter-gatherers arrived. Around 4,500–3,500BP the climate became cooler and wetter, the forest was cut down and at Castlehill, a peat bog began to form on the debris of past glaciers. Conditions on Islay have remained near perfect for peat formation ever since.

We walk past the 100m contour, far above the heights reached by the swollen post-glacial seas and reach Castlehill Moss, 130m above sea level and 4.8km (3 miles) from the sands of Laggan Bay. In the distance, I can just make out the white-walled buildings of Bruichladdich Distillery on the far side of Loch Indaal. I open an unlocked gate and Paddy and I approach a five-metre-high pile of last year's harvested peat. Paddy is not interested, but up close I can see that the dark brown, almost black peat sausages possess veins of tough ancient grasses, sedges and reeds with the occasional imperfection of a tree root, twig or branch. Hard, dry and cracked, these extruded compressed sausages, more bratwurst than pork link, are ready to burn. Tools of the modern-day peat harvest lie scattered around – a shipping container, excavator, two high-end doubled-tyred tractors, large trailers with caterpillar tracks, a potato harvester and a stack of cut timber logs to construct floating roads across a wet bog.

On towards the wind turbine and now at the end of the track, we stop and survey the scene below. Castlehill Moss has been levelled almost perfectly. Ready to harvest, the surface is strewn with fragments of white quartzite. To the eye of a satellite, this 600,000m^2 site looks like a huge earth-brown corrugated roof with parallel drains spaced 20m apart, many water-filled, reflecting white in the remotely sensed imagery.

On our way back down the track, we meet the start of the dayshift firing up the peat harvesting machines – three young contractors working for the company that supplies peat to the maltings at Port Ellen. Despite appearances, they tell me that working the two to three-metre-deep peat can be challenging at times. In places, it reaches a thickness of five to six metres, and on one occasion a digger had to be exhumed after disappearing into one of the wettest and deepest parts of the bog. Large trees are a constant menace and occasionally have to be dug out by hand. They tell me that the harvest has started late this year because of the patchy weather. The boys were 'just getting going', so I leave them to their work. On this clear bright day, with a strong breeze blowing in from the south-east, Castlehill is at the start of its annual peat harvest.

Times have changed on Islay and machine has largely replaced the peat-cutters' tool. Called 'quite possibly the most famous distillery worker in the world',[5] Iain McArthur is a familiar face at Lagavulin Distillery and the possessor of a razor-sharp wit, much to the enjoyment, or occasional bemusement, of its many visitors. He also knows a thing or two about peat and told me, 'When I was a boy, I helped my dad with the peats on Machrie Moss by the airport road when he worked for Laphroaig. In those days, you were paid by the perch and not by time. A perch was five-and-a-half square yards of hand-cut turfs, and you were expected to cut ten to fifteen perches a day'. That is a lot of peat and resulted in long, hard days spent at the peat banks when the weather was fine. Iain began working at Port Ellen Distillery in the 1970s when machines started to be used to cut peat. Towed by a tractor, the first ones 'had a large chainsaw that cut down through the turf, which allowed the wet peat to be sucked up from below and extruded onto the surface where it dried, like sausages'. When the Distillers Company Ltd (DCL, formed in 1877) opened the maltings at Port Ellen in 1973, and after the distillery closed ten years later, Iain got a job at Lagavulin and has worked there ever since. The harvesting operation moved to Castlehill in the late 1980s.

Castlehill Moss surprised me a lot, and despite what I had read and heard about Islay and its famous whisky infused with the flavours of seaweed and past salty, iodine-rich oceans, this is not a coastal bog on the shores of the Atlantic Ocean. It is easy to have a bit of fun with whisky tasting notes, especially those written by the Scotch Malt Whisky Society (SMWS). This one is for a cask of Caol Ila called Firecracker Roll:

Peat cut from Islay contains a multitude of coastal elements including dead shellfish, seaweed, saltwater, and other oceanic minerals that when burned, produce a dense, medicinal smoke that tastes like a salty, briny smack in the face! At nine years old, this whisky is a wild and vibrant spirit with intoxicating plumes of bonfire smoke and smouldering meat. Dampened seaweed and brine are all delivered in a rich and rather intense package that evokes the classic Islay style.[6]

On Castlehill Moss, where this young whisky got its peaty smack in the face from, there is no evidence of an ancient sea. Below the peat lie the boulders, sands and gravels that mark the end of the last Ice Age and not a beach sand or a layer of beautifully rounded pebbles. The flora and fauna of the ocean, now just a distant shimmer, were never part of the story of peat here.

Edinburgh

I am holding a faded dark-blue book, the cover of which is illustrated by various pieces of prehistoric vegetation, amongst them the unmistakable drawing of *Sphenopteris affinis*, the wedge fern. Inside the front cover a handwritten message reads, 'To Mike. 1980'. It is boldly initialled **WRG**, William Reginald Griffiths, my grandfather, a Football Blue at Queens' College Cambridge, wounded in the Suvla Bay landings at Gallipoli in August 1915, decorated soldier and Chaplain to the Forces, awarded the MBE in 1944, Freemason, man of books and Herefordshire vicar. He gave me this small book one year after I started a PhD in Geology and nine years before he died at the age of 93. He felt that *The Story of a Piece of Coal, What It Is, Whence It Comes and Whither It Goes* by Edward A. Martin, FGS[7] would be safe in my hands and a useful addition to my growing library of geology books and research papers. It was, and I treasure it still. Published in 1896, the year of my grandfather's birth, it was one of a short series of pocket-sized informative books, part of *The Library of Useful Stories*. Other members of the series included *The Story of Primitive Man* by Edward Clodd and *The Story of the Solar System, Simply Told for General Readers* by George F. Chambers.

I am told that *The Story of a Piece of Coal* is a classic and my grandfather was almost ceremonial in his giving of the book. It is wonderfully illustrated and includes chapters on the origins of coal and its Carboniferous fossils, the different forms of coal, where it is to be found in the world, a guide to the making of coal gas and even the dangers to miners. Coal is created by the burial and compaction of wood, soil and peat. Over a relatively short period of geological time, this results in the formation of peat-coal. As the impurities are squeezed out by burial and deep geological time, horizontal layers of brown coal or lignite are formed followed by bituminous coal, and finally the most prized and energy rich form of all, grey lustrous anthracite. Sometimes called 'blind coal', anthracite is so pure than when burnt it glows without a flame.

The stories of coal and peat are in some ways inter-linked and in his 1,300-page treatise *Principles of Physical Geology* published in 1944, Arthur Holmes estimated 'that at least a foot of peat is necessary to make an inch of ordinary coal'.[8] Whilst coal is a familiar object to us, peat is not. A tour of most Scottish distilleries is rarely complete without being handed a dry piece of peat, either an odd-shaped nugget, something that looks like a sausage, or a hand-cut turf. For many people, this is the first time they have set eyes on peat, *mòine* or *mona* (Scots and Irish Gaelic), *tourbe* (French), *torf* (German), *turba* or *turfa* (Spanish and Portuguese). As we pass the sample of peat between us, some handle it carefully, almost reverentially; others nose it, disappointed by its lack of smell; some spare it no more than a glance. Look closely, however, and the remarkable story of a piece of peat begins to unfold. In front of me is a specimen of peat that I collected, for science, from a stack of hand-cut, drying turfs on Islay. It can be evaluated and sensed just like a whisky.

Peat – What It Is

General: appearance of a hand-made and hand-sized slab of dense rough-cut organic chocolate cake, with six distinct faces, four of which have clearly been fashioned by a blade. Surprisingly light in weight; a soft, solid brick.

THE STORY OF A PIECE OF PEAT

Colour: dull, brown-black with much lighter orange-brown flecks. Fibrous and non-uniform and characterised by darker layers, with some of the strata opening into air-filled elongated cracks and elliptical voids.

Nose: odourless.

Mouthfeel: rough, grainy or slightly gritty. Tasteless. With water, little change. Most runs quickly away or rests as droplets on the surface – dried peat is hydrophobic, or water-hating. In areas around the large cracks where water is retained, the peat turns a deeper brown colour.

Finish: disappointing. Short.

Rooted in childhood and in the interest of science, I am still occasionally prone to geophagy, the practice of eating soil or clay. It has its roots in superstition, magic and fertility myth, and is still practised in parts of the world today, particularly in Africa and the southern US states, where it is used for medicinal reasons, during pregnancy as a source of iron, as a detox agent, and sometimes as a famine food to alleviate the pain of hunger. I can find no records of people eating peat. I'm sure they did but can reveal it is tasteless.

Look closer still at this deep brown piece of the earth with the aid of a binocular microscope or a hand lens. The light-orange flecks have been transformed into an inter-connected mass of fine rootlets. The larger ones look more like woody twigs or small stems. Burial and pressure from above have flattened them into recognisable layers that now form thin strata. In natural light, small rounded amorphous flecks of clear quartz and a dull white clay mineral become visible. In addition to the cracks and voids, hundreds of minute air-filled pores are now visible.

To get ever closer to the story of a piece of peat, a microscopic, thin section tells us more about its past. It is made by impregnating a wafer-thin slice of peat with resin and allowing it to harden and dry. The peat turns into a piece of rock, which when glued to a transparent glass slide can be carefully ground down to 0.3mm, the thickness of a leaf that can now transmit light. This reveals a new world of ordered plant cell structures, preserved microfossils and identifiable tree pollen, a connected structure of air-filled pores and occasionally a cubic crystal of iron sulphide, 'fool's gold'. Even greater treasures may start to reveal themselves – the hard shell of an ancient beetle or the remains of a *chironimid* – a non-biting midge. Our peat brick has been transformed under the microscope into a nugget of information – a window into the past.

Peat forms from the remains of plants that undergo a long process of natural maturation within the confines of a cold, wet bog. Below ground, the first compounds to break down are simple ones like sugars, followed by cellulose, a major constituent of plant cell walls consisting of flexible chains of joined-up glucose molecules. Their sugary breakdown products ferment naturally and quickly within the bog. Structural components that strengthen, waterproof and protect the cellulose take longer to break down. These include waxes, fats and lignin, its visible brown or orange fibres resistant to decomposition. Peat chemistry changes with time, and over thousands of years compounds disappear and new ones are synthesised. The most abundant chemical constituents of peat are a group called the phenolic compounds of which there are thousands, ranging from small ring structures to enormous macro-molecules. These are also the most important compounds with respect to whisky flavour and are based upon a single (monophenols) or multiples (polyphenols) of a single building block: a six-carbon ring structure with a lone hydroxyl (-OH) functional group attached to the outside. All phenolics belong to a wider group of aromatic compounds that are characterised by ring structures — the word being derived from the Latin word *aroma*, meaning 'sweet odour'. A group of monophenols called phenolic acids create the acidic conditions in peat that inhibit bacterial decomposition below ground. These are sometimes known as the antiseptic organic acids.[8]

One of the most important constituents of peat is lignin, a large stable, non-volatile polyphenol that resides unchanged in peat bogs for thousands of years. At this stage, there is no flavour, smell or aroma associated with these natural phenolic compounds. To stimulate our senses, they need to be combusted, slowly oxidised and broken apart into smaller molecules that are now mobile and volatile. They become part of an array of compounds called congeners – the flavour elements of whisky – that are produced at different stages of the production process.

Peat is the most chemically complex raw material used in the whisky-making process. It is both nitrogen-rich and often sulphur-rich, and while the focus in the whisky world is on phenolics, peat contains other natural components that potentially bring interesting flavours to whisky. The list includes waxes, oils, dyes, tars and fats, all of which were once extracted and manufactured from peat. If the plants that produced the peat were more aromatic and resinous, the resultant peat would contain more oils and fatty constituents.

Although dry peat is odourless, wet or freshly cut peat consists of 90 per cent water and usually gives off a fresh, mossy, damp smell. If you break open a piece of wet peat and are quick enough, it is often possible to pick up the smell of rotten eggs or hydrogen sulphide, a gas produced in the oxygen-starved environment of a peat bog. Sometimes, in places close to the sea, it is even possible to detect a whiff of something that smells iodine-like.

If a piece of dried peat leaves you with a feeling of sensory deprivation, that changes immediately when peat is burnt in a fire grate or kiln. Although I have experienced peat fires in Scotland, Ireland and Finland, it is not easy to describe why they are so evocative, but I will try. In a blackhouse in Lewis, I remember the softness and sweetness of peat smoke drawing me inside through the low front entrance, into an internal blackness. Peat smoke has complexity and depth; it is completely different to a wood fire with its tarry, ashy, resinous and in-your-face, punchy, strong aromas. Or the hot, acrid, khaki-yellow sulphur smokiness of a coal fire that seems to stick to the back of your throat. Comforting, soulful, timeless, even romantic – to me, peat smoke is all of these, and it is the aromatic compounds released by fire that create this unique sensory feeling. It is often said, although I have never seen it, that peat can burn with a soft blue flame. Often mistaken for sulphur, it is the visual expression of bubbles of combustible methane and acetylene gas trapped in voids inside the peat, and only formed in the most oxygen-starved conditions deep inside a bog. Returning to Stornoway on the Isle of Lewis by boat across The Minch, the locals would smell the peat fires of home across the sea long before they reached port.[9]

The peat or peatiness we smell or taste in new-make spirit is therefore the result of a new set of complex organic compounds created by combustion that in turn are modified by mashing, fermentation and distillation. The transformation into whisky that takes place during cask maturation introduces a further, final nuance to these flavour compounds, which had their origin thousands of years ago in a wet, cold, peaty place.

I like this definition of peat from a 1943 wartime pamphlet on the *Peat Deposits of Scotland*:[10] 'an accumulation of more or less decomposed plant remains formed on waterlogged sites, swampy tracts or bogs'. It is suitably ambiguous and gives both a clear feeling of place and a lack of uniformity. In peats close to the surface, those plant remains are fibrous, hairy, often tough to cut, but largely recognisable. Mosses, cotton-grass, reeds or woody material are all important ingredients. The peats are lignin-rich, packed with brown or dark orange fibres that have resisted the processes of decomposition. With time, the fibres and particles become smaller and smaller until they are unrecognisable. The deepest and oldest peat is amorphous, a type of organic clay; rich, wet, heavy and plastic.

The story of a piece of peat starts much in the same way as the story of a piece of coal. In the presence of sunlight, atmospheric carbon dioxide is trapped and photosynthesised by plants to produce organic carbon compounds, carbohydrates such as sugars and starch. Our atmosphere is a rich source of carbon dioxide, currently with an average concentration of 415 parts per million (ppm) or 0.04% and rising.[11] The plants are effectively drawing down carbon dioxide from the atmosphere, harvesting the energy of the sun and transforming it into chemical energy that they store in their living cells and tissues. When plants die, and especially in cold, wet places like bogs and wetlands, the decomposition of carbon-rich plant tissue is slowed down to a point where the dead organic matter begins to accumulate. Given time and the right waterlogged conditions in bogs, mires or marshes, about 1mm of new organic matter will accumulate at the surface each year and the bog will begin to deepen. The story has now begun, but scientists have decreed that it cannot be called peat … yet.

The first recorded academic paper on peat was written in 1685 by William King, Archbishop of Dublin,[12] and published in the *Philosophical Transactions of the Royal Society of London*. Entitled *Of the Bogs, and Loughs of Ireland*, it starts with these words: 'We live in an island almost infamous for Bogs, and yet, I do not remember, that any one has attempted much concerning them.' The Archbishop goes on to describe the 'unwholesome, putrid and stinking vapours' that rose from the bogs, the 'corruption' of the water, 'tinctured by the reddish black colour of the turf'. On preservation, he writes of a piece of 'strangely' preserved leather, about 'bog butter' and the trees he found deep in the wet bogs. A study of its age, at times it deviates wonderfully off-message: 'They are a refuge for *Torys*,[13] and Thieves, who can hardly live without them.' Throughout his paper, the Archbishop is much concerned with the wetness of the land and the need for drainage to make the bogs usable and profitable for turf cutting and animal grazing. With the addition of lime, this was a way of improving fertility or 'curing' a peat bog. In his travels in Ireland, he describes how 'quaky Bogs' mature and grow into 'turf Bogs', an important distinction and a first step in their classification.

Early writings on bogs saw the presence of preserved fallen logs, sticks and rushes as evidence of God's divine judgement, in which he unleashed a flood from the heavens of biblical proportions to purge the earthlings of their sins: 'Various are the conjectures on the origins of bogs: Many suppose them co-eval with the deluge.' Mr Nicholas Turner, who wrote these words in 1784, was clearly sceptical[14] — 'this opinion is liable to many objections' — and from then on scholars began to agree on less catastrophic beginnings and started to develop a language to describe them.

Peat science terminology, like other parts of the scientific lexicon, can sometimes be a bit of an impenetrable boggy mess. Descriptors have evolved at different times and in different places, and an agreed worldwide approach has taken time. Even when soil scientists in the UK tried to decide on a definition of peat, they failed to agree. The soil surveyors of England and Wales differed from their Scottish counterparts when they decided to define peat as a soil with an organic-rich surface layer greater than 40cm, rather than the 50cm favoured by those north of the border. To make matters worse, the FAO (Food and Agriculture Organisation of the United Nations), the creator of the most widely and internationally adopted system of soil classification, preferred to opt out and not call it peat at all. They chose the word 'histosol'[15] (from *histos*, Greek for tissue). Peats in the eyes of the FAO were tissue soils. It's unlikely that the whisky industry will be adopting the FAO terminology any time soon – highly histosol-ed whisky anyone? Yes, it's difficult not to get bogged down in peat terminology. The rather messy, imprecise and unsatisfactory end to this boggy tale is that to be called a peat in Scotland, Finland, Japan or Tasmania, bog-ologists have decreed that there needs to be at least 30–40cm of organic-rich soil above a mineral or rocky base. If there isn't, it's not peat.

Levels of peatiness in whisky are measured in units of ppm phenols, the most robustly peated whiskies like Ardbeg, Laphroaig and Octomore from Islay having the highest values. Peat science also uses a scale, known as the von Post humification scale, to measure the 'peatiness of peat'. The word humification is derived from the Latin *humus* (or earth), basically meaning that the plants are undergoing earthification. Lennart von Post, who devised his scale in 1924 while working for the Soil Survey of Sweden, used it as a measure of peat decomposition.[16] The von Post scale is a wonderfully sensory field test involving giving a handful of wet peat a damn good squeeze. The peat scientist records the colour of the liquid released, then the amount that squidges out between your fingers, and finally notes down any recognisable plant remains that are left behind in the palm of your hand. If nothing really happens after a good squeeze and the plant structure is intact and clear colourless water drops to the ground, a score of H1 is recorded. The Orcadians call this 'foggy' peat, meaning mossy and undecomposed. If the opposite happens, and a black slime-like jelly or ooze rapidly squeezes out through your fingers and then drops on to your boots, it is given a score of H10. The deepest and oldest peats score highest on von Post's scale; they are the densest and most highly decomposed containing little discernible evidence of the plants from which they formed. The Shetlanders call this 'blue peat' and local children would write their names in a material that had the consistency of days-old, cold porridge. It is this peat that is the most carbon-rich and of greatest value as a fuel, the *mòine dubh*, or black peat – the heaviest and darkest. For a modern distillery or maltings, the ideal distillery peat for the production of a peated smoky malt would be rough, fibrous and lignin-rich, something that would smoke and smoulder rather than combust, and score between H4-H6 on the von Post humification scale. Whilst other more scientific and arguably precise methods are now used to measure decomposition in peat, none of them quite match the von Post scale for shear boggy enjoyment.

Peat has been likened to the organic skin of the earth, but in the eyes of both scientists and the whisky industry is much more than a carbon-rich fuel. It is a complex and variable living material that in its natural environment continues to change, grow and mature over time. When burnt, it releases energy and a smoke rich in volatile phenolic compounds, furfurals and hydrocarbons that are vital to the making of wonderful peated malt whisky.

Peat – Whence It Comes

After arriving on HMS *Beagle* in March 1833 from Tierra del Fuego, the year that slavery was abolished in the British Empire, Charles Darwin's first impressions of the Falkland Islands were less than favourable:[17]

> Took a long walk; this side of the Island is very dreary…It is universally covered by a brown, wiry grass, which grows on the peat…The whole landscape from the uniformity of the brown colour, has an air of extreme desolation.

A few days later, the twenty-four-year-old naturalist wrote, 'This is one of the quietest places we have ever been to – not one event has happened during the whole week'. Arriving from the Antarctic a few years after Darwin on board a storm-battered HMS *Erebus*,

Joseph Hooker,[18] the expedition's botanist, wrote to his father, 'Such a wretched place as this you never saw'. In honour of Darwin, the locals named a town after him. After long months at sea, both were obviously hoping to find something more exotic than a group of islands covered by 45 per cent peat – the highest of any country or territory in the world.[19] Its importance to the islanders is celebrated by an annual bank holiday, the first Monday of October – Falklands Peat Cutting Day.

Peat is a truly global soil and while most is found in the cool, wet regions of the northern hemisphere, it occurs in all major climatic regions and on all seven continents. Peatlands cover around 400 million hectares (ha) of the Earth's surface, with the most extensive deposits found in Northern Europe, Russia and Canada. An active and productive peatland acts like a giant carbon sponge, drawing down carbon dioxide from the atmosphere. It is nature's own carbon capture and storage mechanism, and in places like the Flow Country in northern Scotland, an estimated 400 million tonnes of atmospheric carbon[20] has been captured in this way since the end of the last Ice Age. It is the UK's single largest peatland, storing more carbon than all of the nation's forests.

Their names begin to tell the story: the Great Bog, Lost River Peatland, the Faemussach ('stinking' moss), the Silver Flowe. Peat bogs vary in size from the huge expanses of Siberia to small isolated peaty hollows in the Outer Hebrides of Scotland. In 2017, a team of researchers working in the Congo Basin[21] discovered a peatland covering an area of 150,000km^2. It was 10,600 years old and in places had grown to a thickness of six metres. The humid tropical wetlands contain some of the fastest growing peatlands in the world and include examples of the so-called Blue Carbon sinks. In the densely forested coastal mangrove swamps of places like north-east Sumatra and the Ganges Delta, peat can grow at a rate of 20mm per year, twenty times faster than the global average.[22]

PEAT AND WHISKY

In the British Isles, peat started accumulating at the end of the Ice Age on a new land surface scrubbed clean by retreating ice. To imagine what our islands looked like at this time, we need to travel to present-day Svalbard or the high Arctic, treeless places of glaciers and frost-shattered mountains. In the boreal summer, fast-flowing rivers fed by glacial meltwater deposit clay, gravel and broken rock in vast valleys and deltas. Crossing these places full of unconsolidated, liquid mud and rock in the summer is at best difficult and at worst fatal. Splashes of summer colour show that flowering plants, mosses and lichens have begun to arrive and get a foothold. In winter during the long polar nights, the land re-freezes and the rivers and their deltas become eerily silent. This is the treeless landscape of post-glacial Britain – a young landscape that was seasonally frozen or extremely wet, covered with newly formed streams and rivers, gravel banks and moraines, pools and lakes. A landscape upon which peat began to accumulate when the climate became stable and warm enough for plants to grow. Year on year, carbon dioxide was drawn down from the atmosphere, converted to organic plant tissue that in time became the carbon-rich soil that we now call peat.

The geologic base of a bog records this moment in time with a sharp transition from rock or sediment to peat. Over the following thousands of years of peat growth, we see changes as our peatlands retain a precise record of the plants from which they formed. The initial colonisers of the bare rock surfaces, scoured glacial hollows and recently formed mineral wetlands were lichens and mosses followed by woody plants – juniper, pine and birch. In the first part of the Holocene epoch, the period of geological time that started at the end of the great Ice Age, the land surface of much of Scotland would have looked much like the tundra or sparsely forested *taiga* regions of the present-day northern Arctic. With time, plant communities changed as the climate went through phases of cooling

26

and warming. New species appeared, others disappeared, before we finally arrived at the wetter and warmer conditions that currently prevail in our north Atlantic climate.

Much like a book, the peatlands of the British Isles retain a record of these past climatic fluctuations as successional changes in vegetation. At depth, we find treeless layers rich in arctic-alpine vegetation, overlain by bands containing the preserved remains of younger pine, hazel and birch. Walking his estate and digging his peats on the west coast of Scotland more than 200 years ago, Osgood Mackenzie scooped up 'handfuls of hazel-nuts as perfect as the day they dropped off the trees' and 'countless green beetle wings, still glittering in their pristine metallic lustre'.[23] His deepest peats were teaming with preserved life. Unable to call on modern methods of dating the past, he named this bountiful window into geological time 'the good old beetle days'. Much higher up in the youngest, most recently formed peat, we find fewer tree remains, telling us that we have now entered a period of blanket bog formation dominated by mosses, sedges and shrubs, a period that continues to this day.

Vegetation succession through peat-time is recorded in minute detail by millions of tiny preserved grains of pollen with beautiful and intricate shapes, each one characteristic of an individual species of plant. Lennart von Post, fresh from inventing his scale quantifying the peatiness of peat, went on to become the father of modern-day palynology, or pollen analysis, that uses these microscopic spores to recreate the Earth's climatic history. In effect, he was using a vast archive of forensic evidence preserved in peat bogs to unlock the past.

In the British Isles, our oldest peatlands have been dated at 10,000 years BP and first started forming in southerly areas.[22] Being warmer and wetter, they were the first places to become ice-free. As the climate continued to warm and became more temperate, the bog lands began to extend northwards and new areas of peat began to form. Radiocarbon dating of the base of peat bogs records the age of initiation and scientists have measured a cluster of dates around 5–6,000 years BP. A period of rapid peat growth and expansion was underway, a time of optimum conditions for bog formation. At this time, these islands in the North Atlantic had become wrapped in a perfect climate envelope for peat growth. Whilst periods like the Little Ice Age in the 1500s interrupted the warming trend and led to a slowing in growth rate, conditions have remained in the most part favourable for bog formation since the end of the last Ice Age. In addition to plants, other colonisers began arriving, moving north in the warming climate. By 4,000 BP, humans had already cleared much of the outlying Scottish islands of trees and the re-engineering of the landscape had started. There is compelling evidence to suggest that the removal of trees from peatlands by axe or man-induced fire caused peatlands to become wetter and as a result grow faster. Deforestation, cultivation, drainage and farming were now well underway. However, many of the more remote, colder and wetter peatlands remained relatively untouched by man until modern times. This allowed steady peat accumulation over the millennia and the formation of deposits that reached up to 10m deep.

In the flatlands of Scotland and Ireland, extensive peat formation provided both the foundation and a strong geographical focus for the modern-day whisky industry. In Islay on the warm, wet west coast of Scotland, the current climate envelope is near perfect for peat formation, providing the conditions for it to grow faster here than in the relatively cool, dry east. Never far from the sea and warmed by the Gulf Stream, the British Isles benefit from a climate that also provides excellent conditions for growing cereals and delivers endless amounts of rain. Much of Britain is therefore blessed with all the natural attributes and ingredients that are required to make whisky; namely, barley, water and peat.

THE STORY OF A PIECE OF PEAT

It was once said that people living in the north of Scotland were 'as Highland as a Peat',[23] and such was its importance that in his tour of Scotland in the mid-1880s Alfred Barnard, the famous Victorian chronicler of the distilleries of the United Kingdom, wrote often about the quantity and quality of peat.[24] He also understood the importance of place. In total, he visited 116 whisky distilleries in Scotland (as well as twenty-eight in Ireland and four in England) and frequently commented on the local peat mosses, which he enthusiastically described as 'celebrated', 'famous', 'splendid', 'inexhaustible', 'of the highest quality', or 'free of all mineral impregnations'. Peat used to fire the kilns at Ardbeg Distillery had 'an absence of sulphur or other offensive minerals'. This suggests he was well aware of natural differences in peat composition. It was also well known at the time that the mineral-rich ash left in the fire pit of the peat kiln after combustion varied between distilleries, linked to the bog where it was cut and the rock on which it formed.

These differences are well founded, and to scientists, conservationists and commercial users, peats vary from place to place, depth to depth, and often significantly. On the surface, this variation is most clearly reflected in marked differences in the peat-forming plant communities, which leave their signature in the chemical and physical characteristics of the underlying peat. There are also other important factors at play; most important of all is their position in the landscape. At their most fundamental level, peat-forming bogs are separated into two distinct types: rain-fed, that get their water from the atmosphere, and freshwater-fed, that source their water from surface streams or deep groundwater. Add geology, topography, altitude and climate into the mix, and you have the range of factors that combine to create peatland areas, or specific bogs, with their own individual character.

Rain-fed bogs, also known as ombrotrophic bogs, are ecosystems that are commonly acidic, reflecting the natural chemistry of precipitation that typically contains very low concentrations of nutrients like calcium and magnesium. All the plant communities that grow on rain-fed bogs are acidophilic, or acid-loving. Some, like the sundew and butterwort, have developed cunning strategies and traps to acquire the essential nutrients that they need by luring insects to their death. Nutrients can also be acquired from the sea, and in places like coastal Islay or the islands of Orkney, peat chemistry becomes more maritime in character. Particles of shell or sand, wind-blown sea spray and mildly saline rainwater can sometimes reach surprising distances inland – the result is a salty peat reflected in subtle but important differences in the bog plant community.

Large areas of the Scottish Highlands and Islands are covered by this group of rain-fed peatlands that includes raised bogs, peat domes and blanket bogs. These are often the places blessed with high rainfall and cool summer temperatures – a climate that many whisky voyagers will be familiar with. Ireland is of course famous for its peatlands and many fall into this category of rain-fed bogs. One of these – the great Bog of Allen – spreads itself over $1,000 km^2$ in the centre of Ireland. Like its counterparts in Scotland, it has an important place in the history of Irish whiskey.

The second group is the freshwater-fed bogs, also known as minerotrophic bogs. The formal name is a clue to what makes them different – they are more mineral-rich and less acidic ecosystems, typically forming in depressions, and include the basin peats and fenlands found in low-lying and coastal areas. They derive most of their water and nutrients from surface streams draining surrounding hills, deep groundwater sources, or in some places, mineral-rich seawater. Minerotrophic bogs are typically species-rich and diverse, where plants such as the spectacular Grass of Parnassus or bog-star can be found. They were some of the earliest to be exploited and worked out, being located close to humans and their industry.

THE STORY OF A PIECE OF PEAT

In summary, peatlands are organic wetlands, vast stores of water and hydrocarbons. Many important types of plants are involved in their formation producing subtle but important differences in peat character. The key peat formers are the mosses, particularly the *Sphagnum* species, often described as the Super Mosses, with their beautiful tiny, star-shaped leaf structures. They are true ecosystem engineers holding twenty times their own weight in water and creating the conditions in the bog to support the next layer of new top growth. They also absorb nutrients and release acidity, modifying the chemistry of the bog water to suit themselves. Herbaceous plants including grasses, sedges, rushes and heather also play an important role. Another important group of peat-makers are the woody plants, such as pine or birch, which are often extremely well-preserved in peat. Like the tree ferns of the Carboniferous coal deposits, they too provide a link to the climates of our past.

Peat – Whither It Goes

Lying at the Earth's surface, exposed and accessible, peat, unlike coal, has always been a cheap source of fuel. In many places, like the Central Belt of Scotland, peat was in such high demand during the Industrial Revolution that bogs were mined out and completely disappeared. All that now remains are names on a map. In the nineteenth century, peat was fuelling the manufacturing of glass, metal, leather, paper, sugar and chemicals, and peat charcoal was even the preferred choice of culinary fuel for Parisian chefs.[25] Energy production, whisky making, charcoal, bricks, wound dressings, a natural sponge for pollutants, soap, paraffin and tar, polishes and waxes, briquettes, paper – the list of its uses is seemingly endless. Peat was and still is widely used in horticulture as a nutrient-rich, absorbent soil improver, and in the early 1900s was trialled as a bulking agent for distillery pot-ale syrup, probably at Dailuaine.[25] During hard times, it was even smoked as a substitute for tobacco.[9] Peat was either cut and dried by hand or extracted and harvested by machine in large-scale operations akin to strip mining. The fate of peat was to be cut, shredded, squeezed, remoulded and then dried before being carted off to the consumer and manufacturer – an unnatural and undignified end.

Peat is now only used in the whisky industry to create and flavour malted barley, but for a long time it was an important source of heat both to dry the malt and fire the stills. When Alfred Barnard visited Glenmorangie Distillery, the local peat was 'the only fuel used in the establishment'. The legendary Malt Mill micro-distillery built in 1908 beside Lagavulin was still using peat to direct-fire its wash and spirit stills right up to the year of its closure in 1962. The two small stills were modelled on neighbour Laphroaig's and used heavily peated malt, peat-fired stills and Islay's peaty water, the aim being to 'distil a whisky according to the techniques believed to have been used by the pre-industrial Islay distillers', and to produce a better and peatier whisky than Laphroaig. With the classic understatement of an Ìleach, retired Lagavulin stillman 'Big Angus' McAffer called it 'a good dram'. Jim Murray described it as 'more intensely peaty than any other whisky I have ever encountered'.[26,27]

In 1917, a reporter writing for the *Aberdeen Weekly Journal* under the pseudonym Buchan Farmer visited the vast peat mosses of New Pitsligo in the east of Scotland.[28] He describes an industry at its peak, stimulated by the impact of war and a scarcity of wood and coal – quite simply, nothing in a peat bog went to waste:

Mr Godsman took charge of the party at the moss and explained that the company's products consisted of moss litter, moss dust, fishcuring peat, distillery peat, and peat for burning purposes. The moss litter is largely used for bedding horses, etc., and in these times of scarcity of fodder it is a national asset of no little importance. It is made from the mossy fibrous peat and is usually of a light-yellow colour. This quality of moss is mostly found at the surface of the bog and extends down to a depth of about two feet. Fishcuring peat is the light brown peat immediately underlying the moss litter peat, and it is used, as the name implies, in the smoking and curing of white fish. I understand it is now being used to a considerable extent in the kippering of herrings. Under this we get a dark brown peat of a still more fibrous nature, but, unlike the two kinds mentioned, dries quite hard. This class of peat is very lasting, and superior for fuel purposes, but is mostly all used by distillers, and gives the whisky that taste which distinguishes the products of our northern distilleries. Underlying the distillery peat we get the black fuel which is solely used for domestic purposes.

The extraction of peat from a bog and its destruction by combustion or a remake in some new product marks a dramatic and sudden end to the story of a piece of peat. There are, however, other endings. Run a bath on the Hebridean islands of Skye, Lewis or Islay and what often comes out of the tap is water with a rich orangey-brown colour. Your bath water may even contain small particles of peat, which quickly sink to the bottom. You are steeping your limbs in a dilute solution of dissolved organic carbon, a natural breakdown product of peat-rich soils that adds a distinctive colour to surface waters in these areas. Chemically, dissolved organic carbon is a cocktail of complex and large acidic organic compounds, some of which may contains hundreds of thousands of individual carbon atoms. The water is naturally slightly acidic, soft, good for your skin and perfect for making or diluting whisky.

It is the same coloured water that fills ditches, streams, rivers and lochs draining peatlands after heavy rain. As the water table rises, it picks up and washes away the organic acids and peaty particles that lie close to the bog surface. Melting snow has the same effect.

Rainwater and meltwater are acting like a liquid conveyor, a vector to carry away organic carbon from the bog, downslope and ultimately to the sea. The carbon conveyor speeds up and slows down throughout the year as river levels rise and fall, but never stops.

The final act in the story takes us back to the beginning. Fast oxidation by fire, or slow oxidation with the help of sunlight in a loch or a slow-flowing burn, ultimately results in the formation of an inert gas. Discovered in 1764 by the Scottish chemist Joseph Black — 'Candles will not burn in it, Animals that breathe it die' –, it was called 'fixed, fixable Air or Carbonic Acid Gas'.[29] We now know this gas as carbon dioxide, and accompanied by small amounts of methane, this is the path that carbon atoms take when they return to the atmosphere after residing for anything up to 10,000 years on the Earth's surface in a peat bog. They have been locked up and isolated within organic compounds of huge complexity that even now we are still unable to fully characterise chemically.

The story of a piece of peat can begin again, as new atmospheric carbon dioxide becomes a building block for peat in the future – it has reconnected with the Global Carbon Cycle after an extended period of dormancy in a terrestrial peat bog. Not all is lost, however – the maltster and distiller have ensured that the story of peat is locked away in a cask of smoky Lagavulin, Highland Park or Springbank. After twenty years or more slowly maturing in a damp, dark place, the barrel is emptied and the whisky bottled. Now frozen in time, it is an alcoholic preserve of the very essence of the bog from whence it came. The journey from bog to bottle finally ends when the seal is broken and the aromas and flavours of peat smoke are released in your glass. The essence of peat lives on. Alfred Barnard,[24] in one of his greatest turns of phrase, called peat smoke the 'incense of slumbering ages' – you are nosing and tasting a combusted soil that formed thousands of years ago from the plants that grew on a pre-historic, boggy landscape. And that is something quite remarkable.

Two

Peat and Whisky – A Golden Age

'Wi' gude peat-reek my head was light.'
James Duff, poet, 1810[1]

'The best fewell is peat.'

In the summer of 1999, Ken Mackinnon was cutting peat on his croft on the treeless Scottish island of Barra when his *treisgear* hit on something solid 80cm below the surface. After further examination, he unearthed the first of fourteen hard peat bricks, the remains of a fossil pyramid of cut peats dating from the Early Bronze Age.[2] Peat had been used as a fuel since prehistoric times for cooking, lime burning, salt production, charcoal making and even cremation, but by chance Ken had discovered the oldest stack of cut peats known to man. Remarkably, several of the individual turfs preserved the clear impressions of two or three fingers and an accompanying thumbprint. Some are so small and closely spaced they could only have been made by a child. To me, the image of a small girl or boy helping their parents with the annual harvest, stacking freshly cut pieces of island peat on an early summer day 3,500 years ago, is timeless. Peat was the only *fewell* available on Barra, and the survival and wellbeing of this west-coast crofting family and their community over the next twelve months would depend on a successful peat harvest. But, for whatever reason, the cut and dried peats were abandoned on the moss, never brought home, and became sealed in time.

Written just after the period of Viking Rule (900–1200AD), the *Orkneyinga Saga*[3] opens another window into the story. A short chapter entitled *Vikings and Peat* tells us that the Norse Earl of

Orkney, Einar Rognvaldarson (893–946AD), apart from being ugly and one-eyed, was bestowed with the nickname 'Turf-Einar', being 'the first man to dig turf for fuel, firewood being scarce on the islands'. Peat ash found in Stone Age sites shows that it had been used as a fuel in Orkney a long time before the Viking longships arrived, but intriguingly the sagas tell us he did this not on Orkney but on mainland Scotland at a place called Torfnes. It has been assumed that this was Tarbat Ness, a lighthouse promontory on the mainland seventy nautical miles from Orkney. Although it was an area the Vikings frequented regularly, there is no peat at Tarbat Ness or evidence that there once might have been. Torfnes literally means 'turf headland', a name commonly used by the Vikings for areas rich in peat like the Westfjord region of Iceland. The more likely location is further south at Burghead, a place the Norsemen also called Torfness, where two large peat basins lie just inshore from Burghead Bay. This would have been a convenient location to exploit, cut and transport the turf the short distance back to the waiting Viking ships grounded in the nearby bay. In the late ninth century, the Vikings were very active along the Moray coast, and Burghead was fortified and occupied by the Vikings for 200 years.[4] Burghead is now home to one of Diageo's largest maltings and in 2010 they opened Roseisle, a giant new distillery that remarkably looks out across the boglands where the Vikings cut their peats more than 1,000 years ago.

It is not surprising that the oceanic islands of Scotland should yield some of the first evidence of peat use by humans. Fuel was essential to life in a treeless landscape and peat was plentiful. Elsewhere, we know that the Bronze Age beaker people cremated their dead on a peat fire,[5] but in more recent times, the archaeological evidence of peat use by man begins to stack up. Pliny the Elder records its use as a fuel (0–100AD) being harvested by 'maceration',[6] the oldest peat cutting tool was unearthed in East Lothian[5]

(0–200AD), and in York the Romans were cutting peat between 300–400AD. By 750AD, peat in Ireland was being widely used either as a direct source of heat, or to make charcoal.[6] In places, peat was being cut with such vigour that it exposed the bedrock below, effectively scalping the land, leaving the mosses exhausted and gone forever.

Peat occurs widely around the low-lying flooded coastlines of the southern North Sea, and in the 1100s it was already been extensively dug in Flanders to provide fuel for the fast-growing cities of Antwerp, Ghent and Bruges. After the coastal peat was exhausted, the peat cutters moved inland to the neighbouring wet and largely treeless province of Holland.[7] By the 1500s, peat-cutting in the Low Countries was being carried out on such a scale that it resulted in the widespread loss of agricultural land, the creation of open wetlands and even the disappearance of whole villages. A new tool, the *baggerbeugel*, was developed by the peat cutters.[8] In effect a dredging net on a long pole, it was dangled from small flat-bottomed boats to cut peat from deep below the water surface and scoop the turfs up onto deck. New canals were constructed to reach this valuable fuel source that powered the Dutch 'golden age', an unprecedented period of urban growth, industrial and cultural wealth that spanned the seventeenth century. At this time peat from 'the Great Bog of Europe' was widely used as a fuel in the brewing of beer and distillation of spirits, especially Dutch gin,[7] and it's likely that smoky peat flavours would have found their way into *Oude Jeneveer*. This unprecedented period of dependence on peat, captured in Van Gogh's 1883 dark and atmospheric painting *Women on the Peat Moor*,[9] gradually began to wane as coal replaced peat. Man was replaced by machine and by the time the peat-fuel industry eventually died in 1950, 10 per cent of the land area of Holland had been lost to peat extraction. The Dutch golden age of peat was over, but the knowledge and skills learnt in the Low

Countries found their way across the Channel, where Dutch peat engineers set to work draining peatlands across England, eventually reaching as far north as Central Scotland and Aberdeenshire. At the beginning of the twentieth century, the Dutch population in Lancashire and Yorkshire measured in the hundreds, drawn north by the need to drain and harvest the local bogs. In the Highlands and Islands of Scotland, as forests were cleared and fuel wood became scarce, peat increasingly became the main source of energy, and is still cut and used to this day by some of its more remote communities.

In Scotland, the written stories of peat and whisky began to merge in the 1600s with records appearing of 'burnt *aqua vitae*' being made in Morayshire. In 1678, Sir Robert Moray writing about the drying of barley over a kiln stated:

> ... the best fewell is peat. The next charcoal, made of Pit-Coal or Cinders; Heath, Broom and Furzes are naught. If there be not enough of one kind, burn the best first, for that gives the strongest impression, as to the taste.[10]

This was a time when Scotland was still a separate and independent nation and travel and written accounts were scarce. That slowly began to change in 1706–7, when the Acts of Union brought England and Scotland together, 'United into One Kingdom', and this new northern land began to attract curious travellers across the old frontier to explore the wild lochs, open glens and remote islands.

These early travellers journeyed through the Scottish Highlands and Islands by boat, on foot or on horseback. Recording in notebooks and diaries, they sat down by the fire in the evening to write up their observations and thoughts about the local inhabitants and their way of life. The writings of Martin Martin[11] (1695), Thomas Pennant[12] (1772) and Johnson and Boswell[13] (1773) were followed by pioneering engineers and geologists like Joseph Mitchell,[14] John MacCulloch[15] and Sir Archiebald Geikie[16] who were sent to survey and open up this 'unkept land'. The railways had arrived by the time Alfred Barnard made his famous tour of the distilleries of Scotland in the mid-1880s,[17] and he produced not only a detailed and highly descriptive record of each distillery, but also a social history of the land, the people, and their rural way of life. These pioneer travellers made much reference to their encounters with whisky transcending the period when it was distilled privately, illegally or legally. In the nineteenth century, they stayed in basic inns or stagecoach change houses, often no more than a shepherd's cottage, that sold 'excisable liquor' or whisky, accompanied by little more than oatcakes, potatoes and mutton. On the sideboard of private dwellings, guests were presented with 'a bottle of whisky, smuggled of course' before breakfast. After the evening meal, toddy would be mixed from jugs of hot water and milk, bowls of sugar and a bottle of 'mountain dew'.[18] The finest toddy was said to be made from the heavier, peaty malt whiskies produced in the Highlands and Islands.

Martin Martin, a Gaelic speaker and native of Skye, journeyed through the Western Isles under sail around the year 1695. He systematically recorded the culture and natural history of the islands at a time when Scotland was still an independent nation and before the Jacobite Rebellions of the eighteenth century. In the context of this story, he writes about peat, the brewing of ale, *aqua vitae* – its consumption and medicinal powers. It is the literary foundation for all who followed, the first being Thomas Pennant who toured the Lowlands of Scotland and voyaged to the Hebrides in 1769 and again in 1772. He was a factual note-taker who railed against the illicit distillers on Islay where 'a ruinous distillation prevails'. He often recalls poverty, hunger and starvation, but also great hospitality, and took a *deoch an doras*, literally a 'drink at the door', near Loch Broom. He records potatoes being grown on peat bogs and then preserved for the winter by kiln drying, and the locals of Perth mixing their whisky half-and-half with linseed oil. But what interested Pennant most on his travels were the castles, abbeys and antiquities, and he seemed to have a particular fondness for rudely carved stones and statues.

The most famous of those early travellers to Scotland were Johnson and Boswell, who came to the Highlands and Islands of Scotland in 1773 at a time when illicit distilling was widespread. One of Scotland's oldest distilleries, Littlemill on the River Clyde west of Glasgow, had just opened, and on Islay Bowmore was still to be founded six years later. They were quite an odd couple: Dr Samuel Johnson was a man already in his sixties – a famous and somewhat eccentric man of letters; James Boswell a much younger aristocrat who became his loyal companion and biographer. They smelled their first peat fires arriving in Nairn in 1773 after journeying up the east coast of Scotland. This they saw as a sign they were on the 'verge of the Highlands'. On reaching the shores of Loch Ness, they were offered whisky in their 'first Highland hut'; Boswell accepted, Johnson declined. Whisky is mentioned several times in their journals; Boswell describes it being offered as a pre-breakfast drink, 'strong, but not pungent, free from the *empyreumatick* [burnt organic matter] taste and smell'. Although the older and wiser Johnson was abstemious of hard liquor most of the time, when he imbibed, he liked it more than 'English malt brandy'. When they reached the Western Isles, Johnson writes, 'The only *fewel* of the Islands is peat. The wood is all consumed and coal they have not yet found'. On Coll, an island of 'excellent peats', they noted that the whisky was plentiful, there were several stills and the locals made 'more whisky than the inhabitants could consume'. I could find no mention in their journals of encountering whisky on the move. As they wound their way slowly through glens, across mountain passes and along the shorelines of '*loughs*', they must have encountered, greeted, stopped and talked to other riders on horseback with their pack ponies, transporting the occasional small cask of whisky. Maybe at the time it was nothing unusual – the Highlanders prided themselves in their open defiance of the authorities and their tax-raising powers.

In the 1700s and early 1800s, and away from the coal-rich areas of central Scotland, peat was often the only source of energy available to the distilleries of the Highlands and Islands. The distillery manager at Cardhu on Speyside told a visitor in 1832 that 'I use no other fuel but peat, which is a great thing for making good whisky'.[19] The annual peat harvest was an important moment in the distilling year. In late spring and summer when the weather became too warm and the water supply ran low, the distillery workers, sometimes accompanied by their families, would set off for the mosses to cut and dry peat for the following year. On a visit to Talisker Distillery on the island of Skye, Alfred Barnard described 'hardy women busy digging and bringing home this fuel, where they stacked it in an open shed'.

Peat harvesting, unlike coal mining, was dependent on the weather, and there were periods in the late 1700s and early 1800s when a series of long, wet summers not only increased the chances of famine, but also significantly reduced peat cutting and drying. The peat harvest was always at the mercy of the famous Scottish weather.

* * *

'Wi' gude peat-reek my head was light', wrote the Perthshire poet James Duff[1] in 1810, and throughout Scotland it was a golden age for peat and whisky, nowhere more so than in the Highlands and Islands. Regionality has always played a major part in defining Scotland's single malt whiskies, and at its heart lies the use of peat. Nettleton[20] described it as one of the 'orthodox tenets' applied to whisky made in the Highlands, where the malt was dried over a peat fire in contrast to Lowland (and Irish) whisky that was unpeated. These distinctions go much further back to the end of the eighteenth century. Frederick Maclagan,[21] an 'Agent and Traveller' working for Messrs Haig at Lochrin Distillery in Edinburgh, said the difference between whisky made in the Highlands and the 'Low Country' was the 'Flavour of the Smoke of Peat', and:

> Whifkey not tinctured with it was reprobated as unpalatable and unwholefome. I have fold Whifkey in Perth tinctured with the Flavour of Peat Smoke, manufactured from Malt mixt with Raw Grain, which was retailed, approved of, and paid for as Highland Whifkey.

He said that one of his salesmen profited more by bringing it to his customers by night:

> … to perfuade them it was fmuggled … It is certain, to the South of Edinburgh the People will not drink Whifkey tinctured with the Smoke of Peat.

This was the time when the flavour of peat smoke became firmly embedded in the DNA of Scotch whisky. It is often described as a moment of serendipity, the occurrence and development of events by chance in a happy or beneficial way. Or was it, as Dave Broom has suggested, a 'conscious, quality-led decision by the seventeenth-century brewers and distillers, who understood the aromas created by the burning of peat'. Whichever way, an unbreakable bond was created between peat and whisky that we celebrate to this day.

*'Ach, I believe you'll have distilled it yourself Doctor,
in a Poit Dhubh.'* [18]

The bond that connects peat with whisky is not only about 'peat reek', but about place; illicit stills were located in a 'commanding eminence', 'in the centre of a bog', or 'in a well secured fastness'.[17] They were also well hidden – 'Once when shooting I fell through the cunningly concealed roof of such a cave into a heap of malt' – the words of Dr John MacKenzie of Eileanach, describing his amusing disappearance through the roof of a whisky bothy, apparently 'a common experience for Highland sportsmen.'[22] The nearest peaty burn or bog pool was not only used as a source of plentiful fresh clean water but as a 'steeping vessel', where sacks of barley were placed for a few days to soften and break the dormancy of the grain. A flowing stream also doubled up as a convenient way of disposing of waste. The barley was then germinated and turned for seven to ten days before being transferred to a peat-fired kiln for drying. After milling, the malt was mashed in the *poit dhubh* – the black pot – over a peat fire, fermented and then double-distilled.[23]

The mythical status of the *poit dhubh* was firmly entrenched in whisky lore long before Compton Mackenzie wrote in *Whisky Galore*, 'Ach, I believe you'll have distilled it yourself Doctor, in a *Poit Dhubh*'.[23] Part of the folklore of the smugglers' bothy, the *poit dhubh* became associated with a seemingly bottomless cauldron of glorious whisky fiction and romance. Suspended above a smoking peat fire, a mash of malted barley was boiled then fermented and distilled. This resulted in the production of 'peat reek', defined in Jamieson's *An Etymological Dictionary of the Scottish Language* (1880) in several ways: 1. the smoke of turf-fuel, 2. the flavour communicated to *aqua vitae* in consequence of it being distilled by means of turf-fuel, and 3. Highland whisky.[24]

Distillation was a Celtic craft and a knowledge of it most likely arrived in Islay from Ireland around the year 1295 with the

Beatons,[25] part of a Royal entourage attached to an Irish princess who came to marry the Lord of the Isles. They went on to become hereditary physicians to the kings of Scotland, including Robert the Bruce, and would have had great influence on medical matters and treatments both inside and outside the royal court. Although monks were already distilling spirits on the island of Ireland in the sixth century, it was the dissolution of the monasteries by Henry VIII in the 1500s that unleashed a wave of knowledge that dispersed across Ireland and found a home in bothies, farms and country estates. Scotland had recorded the first use of *aqua vitae* in 1494 at Lindores Abbey in Fife, and the seeds were sown for the widespread making of *uisge beatha*. Over almost three centuries until 1779, private distilling in stills of less than ten gallons was legal, but that year the still size was reduced to two gallons, and then in 1781 the practice was made illegal. This triggered a time of smuggling, evasion, ingenuity and hardship in places that became future centres of legal distillation. At this time, whisky making in Ireland and Scotland was often closely linked, promoting the free movement of people, skills and whisky across the narrow stretch of water that separates the two countries.[26] The links clearly go back further; Martin Martin,[11] travelling in the Western Isles a few years before Scotland became part of the Union observed that the natives of Islay and other islands 'generally speak the Irish tongue'. As a speaker of Scots Gaelic, he would have known.

The Kingdom of Ireland was ruled by the monarchs of England and Great Britain from Dublin Castle between 1542 and 1800, when it became part of the United Kingdom of Great Britain and Ireland. That lasted until the formation of the Irish Free State in 1922. Ireland was and still is a country with huge peat resources, and at one time its whiskey or *poitín* – meaning 'little pot' – would have been as peated as Scotch whisky. At the end of eighteenth century, it is estimated that there were 2,000 mostly illegal stills in Ireland,[27] typically concealed in small bothies and largely dependent on peat for fuel. Illicit *poitín* was traditionally made 'beyond the pale',[28] essentially a large ditch or barrier that surrounded the crown-controlled lands of Dublin. Beyond the pale was poor Gaelic-speaking Ireland with its peat-rich lands, illicit *poitín*-making stills and smugglers' bothies. *Poitín* production peaked in the late 1700s and early 1800s, a period that became known as the '*poitín* years'.[29] Peat use began to die out in the nineteenth century, and eventually the open, smoky peat fires of the illicit still were replaced by modern, small and portable smokeless gas burners.

A visit to a smugglers' bothy in the hills of Northern Ireland in the early 1800s by Michael Donavan, a professor of chemistry, provides one of the most vivid descriptions of the home of a *poitín* maker.[30] He writes of a 'very small thatched cabin with a large turf fire' and an opening in the roof to carry away smoke that was 'so acrimonious that my eyes were exceedingly smarted'. The fumes 'consisting chiefly of pyroligneous acid in vapour' were the result of four distillations that produced what he called wood vinegar or wood acid from the combustion of lignin. The bothies of the *poitín* makers, clearly of interest to a chemist, were dark, damp places of work, and peat smoke impregnated the very fabric of the dwelling. Frederick Maclagan again recorded the prevalence of peat use in whisky making in the Highlands of Scotland, saying that it was:

> ... not in their [the Highlander's] Power to manufacture it without that Flavour, not merely on Account of their drying their Malt with Peats, but from their constant Ufe of that Species of Fuel in their Families, which gives the fame Flavour to their Milk, their Butter, their Cheefe, their Meal and Bread, even their wooden Utenfils and very Clothes being affected with it.

The tipping point for the *poitín* and the whisky smugglers of rural Ireland and Scotland was the Excise Act of 1823. From that point on, illicit whisky gradually became replaced by licit whisky made firstly in small licensed farm distilleries that with time were succeeded by larger commercial enterprises. But the making and trafficking of illegal whisky was so deeply rooted in rural culture that the demise of the illicit still changed the social fabric forever. Almost 200 years later, the places and people live on and their stories lie unburied and just below the surface.

Peatfold, Glenbuchat, Scottish Highlands

It's May and more than thirty years since I last set foot in Peatfold. A cold wind is blowing out of the north and a fresh dusting of snow covers the rocky tors that cap the surrounding Glenbuchat hills. A May day that calls for warm clothes, but the sounds and smells speak of spring. The curlew and lapwing have returned to the glen, now paired off for the breeding season; their sharp high-pitched calls ring out as they wheel across the treeless, open moorland. Gorse is in full bloom and up close, the bright yellow flower clusters smell of warm coconut milk. Maybe the grass and sheep are more abundant than they were back in the 1980s, but to me the glen seems to have changed little. Fertile arable fields, linear blocks of tall forestry plantations, abundant juniper and still a profusion of wild rabbits. And higher up, a patchwork mosaic of burnt heather of different ages and colours, providing food and shelter for red grouse.

The solitary scraggy trees are still there and the same peat-rich burn comes tumbling downhill next to the track, but Peatfold itself has changed. The building has been completely renovated and two shiny satellite dishes now adorn the walls of the cottage. At the front, there is a neatly cut lawn; at the rear, wet clothes are drying in the cold air. Back in the 1980s when I worked here, it was an unoccupied, derelict farm cottage. It was one of five buildings that in 1850 had made up a farmstead located next to the whisky smugglers' path that linked the Cabrach to Strathdon and the markets in the towns and big cities beyond. Back then, it had a water-powered threshing machine and enough roof space to accommodate local shoemakers and tailors.[31] Fuel for the farmstead came from the local forests and a peatfold, a nearby lobe of peat perched 400m uphill on a plateau that still bears the scars of the peat-cutting tool.

In 1850, Peatfold was home to Isabella and Peter Gow and their family. 'Bell' was born in 1791 in Glenfeshie, a tributary of the River Spey, one of fourteen children who comfortably outlived them all. Her father, Allan MacDonald, died when she was three years old. He worked as a boatman, floating timber logs down the River Spey. When Bell was a young girl, she herded the small black Highland hill cattle, worked in domestic service and 'assisted a good deal in the smuggling of whisky, and could tell of some rare encounters with the gaugers'. Bell's husband Peter was born in Glenbuchat in 1794. From an early age, he was part-time whisky smuggler, part-time crofter. Bell married Peter Gow over the hill in Cabrach in June 1823, the same year that the Excise Act was passed when the tax on spirits was reduced by 50 per cent and a minimum still size set. They lived in Peatfold next to the smugglers' path and raised a family of six. Peter died at Peatfold in the cold January of 1857. Bell lived on to be 104 before finally succumbing to influenza in 1894.[32]

The smugglers' track to the Cabrach climbs steadily north from Peatfold and follows the burn across the treeless moorland. The characteristic smell of charred heather still hangs in the air – aromatic and sweet, burnt hot leather, a bonfire on the hill. Higher up, I hear the crack of a single rifle shot – a gamekeeper at work. A network of rough gravel paths and vehicle tracks criss-cross the hills and disappear over high passes into hidden valleys. Beside the path, I come across a cairn, a large rough-cut pink granite grave-stone surrounded by a carefully stacked halo of angular white quartz boulders. Chiselled into the granite is the inscription *A DAVIDSON 1792–1843*.

Alexander 'Sandy' Davidson, 'poacher, smuggler and social out-cast', was a colourful character who wandered far and wide in the forests and hills of north-east Scotland in pursuit of both lawful and unlawful practices.[31] For a number of years, he was a key member of a smugglers' gang working out of the hills near Braemar and Glenlivet, carrying whisky to market in Aberdeen on the backs of the Highland shelties – the small tough ponies much favoured by the smugglers. A rare image of Sandy shows a smartly-dressed kilted man in his thirties with a tanned face, sharp nose, dark well-kept beard and moustache. He is holding, almost cradling in both hands, a rifle – the Scottish fur trapper at work in the Canadian Rockies – a solitary John Muir of the forests, mountains and wide-open spaces. Always his own person and deeply religious, he was a staunch defender of the Sabbath. Sandy was also a deadly shot, a skilled stalker and, it is said, an elegant dancer. He died of natural causes at this cairn-marked spot on the hillside above Glenbuchat, discovered by an August shooting party. Charlie, his little brown pointer, was found sitting on his chest protecting his by then cold and stiff master.

The personal stories of Isabella Gow, Peter Gow and Sandy Davidson tell of a time when Glenbuchat and its people were a focal point for the making and trafficking of illicit whisky. During the first twenty to thirty years of their lives, illicit distilling and the smuggling of its produce were very much part of the social fabric of rural Scotland, especially the Highlands. Five miles north of Glenbuchat lies the Cabrach, and eleven miles to the north-west the fertile green fields of Glenlivet in the Spey valley, both well-es-tablished meccas for the illicit still in the late eighteenth century.

Even in 1819, John Smith, the smuggler and future owner of Glenlivet, recalling a raid by excisemen on the glen described how 'the gauger's eyes were rejoiced to see some forty or fifty smugglers' stills reeking away finely'.[33] Glenbuchat was located at a crossroads where pack trains of ponies on the old road coming down from the high hill passes to the west met and mixed with the southbound whisky smugglers from the Cabrach.

In its heyday in the early 1800s, illicit distilling in the Cabrach was pursued on an industrial scale with over 100 small stills housed in hill farms, bothies or houses.[33] Their produce fed the smugglers' roads, taking illegal whisky south carried on the backs of their ponies in ten-gallon ankers. The whisky had a fine reputation and was destined for the cities and towns of the east coast of Scotland. Nowadays, the Cabrach is still a wild place of rolling moorlands of heather and juniper, with small patches of arable land on the better more sheltered soils. Peat is ubiquitous in the area and peat fuelled the stills of the Cabrach. Juniper wood, a smokeless fuel much favoured by the illicit whisky makers to avoid detection, was also burnt in the stills by William Grant & Son to cleanse and sweeten the pot. Famously, an illicit still was even found concealed inside a peat hag.[34] After the 1823 Excise Act, three small-scale legal farm distilleries started up in the area. They operated for a few years before closing in the 1840s. While the great days of whisky making in the Cabrach were seemingly over, peats from the mosses continued to be cut well into the twentieth century for the local Dufftown distilleries of Balvenie and Mortlach.

Although not on the scale of the Cabrach, Glenbuchat was a neighbouring centre for illicit distilling and was famed for the quality of its whisky, said to be as good as the Glenlivet from over the hill. Barley grew well in the glen, local lime kilns produced quick lime to improve soil fertility, water was plentiful and names on the map like Peat Hill and Moss Hill tell of local peat sources. It had its own smithy and a mill with a drying kiln.

Like the rest of the Scottish Highlands, the tipping point for the illicit still in Glenbuchat was the year 1823, a year that triggered huge changes to the lives of people like the Gows and Sandy Davidson. The area became a regular target for the excisemen, or gaugers as they were often known. In 1821, a raiding party searching for illicit stills took away and charged thirty-nine Glenbuchat men, resulting in jail sentences for some of the convicted.[31] Corgarff Castle, just 13km (8 miles) away, was built as a military garrison and watch post for whisky smugglers travelling along one of General Wade's new military roads. In a further effort to control the illicit trade, excisemen lived in rented property in the glen. James William Barclay, land reformer and local MP who bought the estate in 1901, said smuggling and whisky making was once 'the most extensive and profitable occupation in the Glen, after farming'.[35]

It was widely recognised that the whisky trade had a negative effect on agricultural improvement and productivity, particularly as jail for the men produced real hardship for the families and took workers away from the fields.[31] However, after the 1823 Excise Act and with a growing temperance movement, the production of illicit whisky in the Cabrach declined rapidly. The minister of Glenbuchat wrote in the 1840s about the improvement to his flock. The population was growing, he said:

> ... in consequence of the increase of cleanliness of the people, greater attention to children in extreme infancy, vaccination, but, above all, the annihilation of smuggling. The improvements in every respect, since illicit distillation has been happily put down, are truly astonishing.[36]

Religious leaders were to the fore in Scotland at this time and their words were often powerful deterrents against the making of illicit whisky. In 1843, in an event known simply as The Disruption, 450 ministers broke away from the Church of Scotland to form the Free Church of Scotland and throughout the land in places like Glenbuchat new churches were built to watch over their people. After the population of the glen peaked in the 1870s with almost 600 residents, in common with many parts of the Scottish Highlands, people started leaving for the growing cities. In 2001, the local population had declined to just sixty.[31]

On my way back down the hill to Peatfold, I meet two young gamekeepers in their green ex-army Land Rovers. They were probably not born when I first came here in the 1980s. One of them was the current occupant of Peatfold and owner of the clothes on the washing line. I told him I had sampled and analysed his drinking water many years ago, and asked him what is was like. 'Aye, pretty good,' he replied cheerfully. 'Nae doing me too much harm.'

I leave Peatfold and its stories of the past and head north past Corgarff Castle along the old military road over the hills towards Tomintoul. My destination is the fertile green valley of the River Livet, but on the way I stop by a place that was very much part of the story of the old Glenlivet, 'the only liquor fit for a gentleman', according to Sir Walter Scott.[37]

The Far-Famed Moss of Faemussach

'The famous Faemussach Moss, with its inexhaustible peat deposits, contributes something to the distinctive flavour of Glenlivet whisky.'[38]

Almost 150 years ago, a group of wanderers 'entered the far-famed moss of Faemussach, a gloomy wilderness of peat'.[39] On a bright spring day I too enter, even wander, into the far-famed moss through a lichen-coated wooden gate to follow a sunken, puddled track into this huge featureless bog. Almost immediately, I come across a clue to its past, the remains of an old peat railway: rotten wooden sleepers, rusting iron rails and soon after, the discarded shell of a peat wagon, its rusting iron wheels partly submerged in a pool of soft rushes.

This area of the moss had been scalped of its peat long ago. A wonderful old black-and-white photograph taken in the 1950s shows a gentleman with a cloth cap, sleeves rolled up, carefully stacking Faemussach peats in the sun.[40] The stacks are small, each made up of ten neatly cut turfs, and he stoops over one of several hundred in a field of recently harvested peat. In the old days, a deep peat moss like Faemussach would be dug over three times in its lifetime, the cutters taking away one layer after another until they reached the base up to ten meters below the original surface.

At one time, the Faemussach was a vital source of peat for the local population of Tomintoul. 'Peats are cut in May and June and left all of the summer to dry. They lie in stacks like sheaves on a harvest field … till September when they are dry enough to be taken in.' The whole town was involved, and reporters liked to paint an idyllic picture of the annual peat harvest: 'Though the work is not light, and they are exposed to any sort of weather that may come, the workers seem to enjoy it. Up on the open moorland the keen wind brings a brighter colour to the cheeks of rustic beauty and heightens the spirits and inspires the gallantry of stalwart young men.' At the end of the day, 'the carts come home across the moor, and soon peat stacks at cottage gables and backyards grow bigger and bigger till there is enough to last the winter'. The Faemussach was also the source of the 'immense quantities of peats taken from the moss for the Glenlivet Distillery, for it is peat reek that gives to Glenlivet the pleasing flavour which – according to those who have tried it – this famous whisky possesses'.[41]

I am no longer a stalwart young man, and crossing the moss today on foot was far from idyllic. It was a struggle, an undulating low-level assault course. Animal tracks give occasional hope of a trail, only for those hopes to be dashed amongst overgrown cutover bogginess. I scramble up old two- to three-metre-high man-made peat banks onto dry land and then jump down again, trying hard to avoid the unseen wettest places below. In the distance, I home in on my target, a large yellow object, a modern, man-made machine.

Although much of the peat is gone, nature is gradually reclaiming the Faemussach. Grasses, sedges and reeds now grow on the nutrient-rich glacial deposits that lay below the peat. It is said that flint arrowheads were regularly recovered from this layer, evidence that man may have lived and hunted here long before the peat bogs formed. A new community of plants has replaced the acidic-loving *Sphagnum* that built the bulky structure of the ancient moss, including yellow buttercups and pink flowering ragged robin that add a welcome splash of colour. Young opportunistic spruce seedlings are beginning to invade the space. Lapwings and black headed gulls have arrived, benefiting from the new fertility and life. Two sika deer lope across the bright green grass. Juniper scrub grows well on the dry raised peat hags left behind by the peat workers and so does cotton-grass, their small white seed heads nodding gently in the wind, primed for dispersal this autumn, a shifting sea of white candyfloss. *Eriophorum angustifolium* is actually a sedge, not a grass, that has become a hallmark of dry and drained bogs. Using the shelter of naturally regenerated trees and bushes, a solitary roe deer crosses in the distance heading north towards Glenlivet, just 8km (5 miles) away.

I close in on my target. It sits on a huge pedestal of black peat and between me and the yellow machine is an obstacle, a four-metre-high peat face exposing a section through time. Within one metre of the surface are the characteristic red-brown roots and branches of Scots pine – judging by the size of the root systems, some of these ancient conifers must have been absolutely huge.

Much deeper, at three metres depth, marking a time when decid-uous trees were dominant, is the unmistakable lustrous bark of silver birch. I scramble up the fragile peat bank and emerge onto a sea of blackness – a raised pedestal of fresh organic matter above the old cutaway bog.

I am now on the surface of an opencast peat mine, stripped away layer by layer just like a surface coal mine. Faemussach peat is fibrous and layered with characteristic curved lenticles that peel away like onion skins when dry. Today, the peat is loose and friable, and when the wind blows the fibres disappear as airborne particles of dust. Closer up, it contains lighter orange-brown flecks of colour, the visible lignin-rich stalks and stems of dead plants. Faemussach peat was regarded as particularly pure, *Sphagnum*-rich and easy to work because of lack of wood and being 'free from all mineral impregnations'.[38] Like many Highland peats, it is famously light and friable, unlike the heavier and wetter Islay peats.

In the middle of this flat, 19ha balloon-shaped platform of exposed bare peat is an abandoned yellow Hymac digger, now more rust than paint. Its giant hydraulic arm with a large artic-ulated scoop rests upright and stationary on the peat surface. As I approach, the air is full of the unmistakable smell of oil reek – leaking hydraulic fluid is leaving an oily sheen on the bog surface. In the same way that tractors were modified to cross soft snow in the golden years of Antarctic exploration, this Hymac, with its two wide wooden caterpillar tracks, had been adapted for working on wet peat. Like an ancient dinosaur's footprints fossilised in Jurassic sandstone, the last journey of the Hymac can be traced by its track marks still preserved on the peat surface. A slow wide turn to the east before straightening up and coming to a shuddering halt in the middle of its four-metre-high-pedestal of peat. Did it break down one final time and become too expensive to repair? Had the demand for Faemussach peat just ebbed away and the buyers gone elsewhere? The yellow machine had now begun to tilt and sink slowly into the bog, its back end weighed down by the mass of its rusting engine like a ship sinking stern first into the sea. The bog is slowly reclaiming its assailant. Close by, I find its accomplice – an abandoned double-tyred, rusty red tractor.

I watch in silence as a solitary hare crosses the bare black plateau, clearly aware and nervous of its visibility. It pauses halfway across in the temporary sanctuary of a tuft of new grass and, thinking better of it, quickly moves on, lolloping across the open ground to seek safety on the far side where it disappears completely from view into new growth vegetation. A rare moment of tension has passed on the bog.

Now on the southern edge of the Faemussach, I walk along a tree graveyard of Scots pine, the grey and white bleached remains of roots and trunks pulled out of the bog and cast to one side. Many bear the scars and claw marks of the machines that ripped them from their home. In days gone by, the ancient wood would have been split into long rods to make flaming smoky torches or stave-sized laths to bring light and life to a peat fire.[16] In Morayshire, large splinters of bog pine were fixed to metal stands to make what were locally known as '*puir men*'. In Speyside, they were called '*fir can'les*' and the resin-impregnated pine produced an 'astonishing blaze'.[39] They were so numerous in the peat mosses around Glenlivet that the locals used the buried fossilised wood to roof their houses.

A new population of Sitka spruce, the size of small Christmas trees, now grows amongst the graves of their coniferous ancestors, the windblown seed germinating in the shelter of the ancient roots. Across the flat bogland and behind the rusting yellow Hymac, Ben Rinnes appears in distant bright sunshine to the north-east. As the breeze suddenly drops, the midges lift off, not biting yet as the season is too early, just irritating. These are the males that only need plant nectar to feed. The females come later in the season hungry for animal nectar, particularly the hairless, soft skinned human variety. A warm blood meal to grow their eggs and impregnate the bog, ready for next year's fiesta.

This recently worked black expanse of peat just outside Tomintoul on the high road to the Glenlivet Distillery is just a small part of the *Feith Musach*, the 'filthy myre'. It is all that remains of the Faemussach peat operation, which once covered an area at least six times this size. For 200 years, the whisky makers of Glenlivet depended on the Faemussach as a source of peat, and when George Smith took out a distilling licence after the 1823 Excise Act and the size of the distillery began to grow, so did the importance of the Faemussach. Several other distilleries also used Faemussach peat including Tomintoul, Springbank and Benriach, the latter having a licence to cut peat at Faemussach during the mid-1980s. The licence for cutting peat expires in 2026,[5] but when I visited the site in 2019, it had already been abandoned. It looked like the will to cut peat here had also expired. There were no locks on the gates or signs proclaiming ownership – two years later, the yellow Hymac had vanished.

The bond between peat and whisky, highlighted by the words 'the convenient proximity of a peat bog is an economic necessity for a Highland malt distillery',[38] is nowhere better illustrated than here. Although Faemussach peat is no longer used to produce Glenlivet whisky, the link lives on in the landscape and in the peat-rich distillery water sources. Glenlivet, like most Speysiders, is now an unpeated whisky. The floor maltings closed in 1966, and that marked the beginning of the end of the bond with the filthy myre. Ex-Islay casks are still sometimes used to generate a degree of peaty smokiness during maturation and in 2008, Glenlivet released an expression called Faemussach, a 13-year-old single malt whisky matured in a second-fill sherry butt. Rather disappointingly, the tasting notes suggested there was not a trace of peat or a hint of smoke.[42]

Few distilleries epitomise the changes that have taken place over the last 200 years of whisky making in Scotland better than Glenlivet. From the time of the *sma' stell* and illegal distilling, it became a legal farm-sized distillery called Minmore, before changing its name, expanding and ultimately becoming the largest producer of single malt whisky in Scotland. It now has a capacity to make an astonishing 21 million litres of pure alcohol (LPA) a year and is home to the biggest aluminium mash tun I have ever seen. Yet despite its size, it still retains the feel of a distinctive place, linked to the past and surrounded by those fertile green fields where barley once grew in abundance.

Times have changed not just at The Glenlivet but throughout Speyside. A 1952 whisky produced at the Glen Grant Distillery was once described by a well-known whisky connoisseur as the peatiest he had ever tasted from the distillery.[27] The malt specification for its mash is now less than 1ppm phenols. In the late 1880s, peat was still being widely dug locally, and distilleries like Glenlossie, Glen Grant, Glenrothes, Aberlour, Mortlach and others all had large peat sheds that were filled to the roof in the autumn.[17] At the time, the malt kilns of the Speyside distilleries were commonly fired by peat or a combination of peat, coke and coal. Its reputation as a one-time peated whisky was widely known, and the taste of whisky produced at the Pulteney Distillery on the Caithness coast was once described as 'never quite so peaty as some of the Speyside stills'.[43]

Peat use by the Speyside distilleries gradually declined with the arrival of coal in the 1860s and the taste of its whisky changed. There was a brief renaissance in the years following World War II as a shortage of coal in Scotland led to a surge in the use of peat. An old black-and-white photograph taken in 1965 at the Macallan Distillery shows both anthracite and peat being used to fire the malt kilns and add flavour to the whisky.[44] In July 2018, a very rare 52-year-old 'curiously peated' Macallan, distilled in 1946, turned up at auction and sold for £14,500.[45,46] A curiosity maybe, but a whisky produced out of necessity at the time.

I asked Angus MacRaild, an expert in old-style whisky, about these long gone peated Speysiders:

I have tasted Longmorn, Glen Grant, Mortlach, Glenfarclas and Macallan from the 1940–60 period and indeed they were often robustly peated and at the time an important part of the taste profile of Speyside single malts. The early wartime years and period immediately post-war when coal was in short supply saw some distinctly peated examples being produced. These expressed typical old-style peat flavours – richly earthy, herbal and with deep phenolic flavours. These malts often tended to display a slightly sweeter peat smoke flavour as well. Glenlivet Distillery also possessed a notable peat character in many bottlings distilled throughout the 1950s, often showing a lighter peat flavour with peppery, waxy and medicinal flavours. These whiskies are all markedly different from today's products, even – perhaps especially – when peated batches are produced at these distilleries today.

In the late 1960s and early 1970s, floor maltings virtually disappeared from the Speyside distilleries, and with this came a marked decline in the use of peat. But the regional character of the peated Speysiders continued to live on in the mind, and even in 1987 their hallmark style was being described as 'smoky, firm-bodied, with more than a hint of sherryish sweetness.'[47] Benriach and Balvenie are now the only remaining distilleries on Speyside that still occasionally operate their own floor maltings. In 2002, Balvenie started malting peated barley again on an annual basis and the Balvenie

Peat Week is a welcome return to a style of whisky produced fifty years ago. Significant amounts of peated whisky are also made today in the distilleries of Benromach and Tomintoul, but none of the peat is harvested locally in Speyside. The peat reek that drifts up to the town of Dufftown from the Balvenie malt kiln for one week each year is the smell of smouldering St Fergus peat, harvested from a moss fifty miles distant on the east coast of Scotland.

The Peat Road

In modern times, Speyside has become well known for a lighter and sweeter style of whisky exemplified by the use of sherry casks, and is home to some of the big hitters like Macallan, Tamdhu, Aberlour and Glenrothes. The region lies at the heart of Scotch whisky production with around 50 per cent of all working distilleries located in the valley of the River Spey and its tributaries. It is easy to see why whisky making here was so attractive to the early distillers. The valleys were wild and remote, there was abundant fresh water, much of it tinged with a brown peaty colour, and local barley was grown on the fertile soils of the river flood plains. On the hills above the Spey, Fiddich and Livet, there was abundant peat and most importantly, the locals knew how to make very good whisky.

Up on the Mannoch Hills to the west of the River Spey is an old peat-cutters' track called the Mannoch Road. Scotland once had many such peat roads, but this one is special. It heads north in a straight line from Knockando towards Elgin and although these days it is no more than a rarely frequented hiking or mountain bike trail, it connects the working distillery clusters of Tamdhu-Cardhu-Knockando with Glenlossie-Mannochmore-Miltonduff, and halfway along its route lies the Birnie Moss. The Mannoch Road dates back to the late 1600s, and legend has it that Helen Cummings walked barefoot each week along this road from Cardow Farm to Elgin carrying illicit whisky made at her farm to sell in town.[48] In 1824, Helen and her husband John took out a distilling licence and founded Cardow Distillery, one of the first in Speyside to do so after the passing of the 1823 Excise Act.[48] In the early years, peat was the only source of energy at the distillery, but that all changed in the late nineteenth century with the arrival of the railway and the coming of coal.

In the 1880s, Cardow, or Cardhu Distillery as it had become known, possessed a spacious peat store with hundreds of tonnes of 'rich odoriferous' peats harvested each year from the Mannoch Hill.[17] As well as its peats, the hills provided the distillery with water in abundance. Like so many Speyside distilleries, Cardhu has lost its link to the past – its floor maltings closed in 1966 and the old peat shed became a boiler house. Modern Cardhu single malt whisky, a key component of the Johnnie Walker blend, is now unpeated, or as 'clean as a whistle', as a well-known whisky writer once put it.[27]

From Cardhu Distillery, I follow in the footsteps of Helen Cummings and her illicit whisky and walk uphill beside the brown peaty source stream, past the man-made holding dam and on into the forest. The old road passes the derelict Mannoch Cottage and carries on rising gently up to the 300m contour. In the summer, this was an easy early morning ascent for the distillery workers, freed from their normal duties in the quiet season. I pass through dense areas of coniferous plantation, recent clear-fells and a large fenced-off area enclosing a young population of naturally regenerating Scots pine. To the distillery workers fifty years ago, it would have looked quite different. In the shelter of the forest, the air is almost still and a pleasant smell periodically drifts up through the trees, the fruity, estery odour of fermenting barley from one of the distilleries in the Spey valley below.

At the highest point on the track, the path suddenly dissolves into a damp, boggy depression and I stumble upon the unmistakable signs of an abandoned small peat-working: overgrown

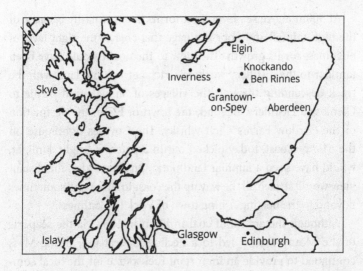

cut-moss banks, bare black peat and ancient pinewood, dragged out of the bog and cast aside. To the west across open heather moorland lies the Mannoch Hill, a gentle, whaleback summit of 'hard Highland Schist'. The old Peat Road then suddenly dives into a blanket of young Sitka spruce, the spiky needles of its branches blocking the path and piercing the bare skin of my arms as I push my way through. This part of the Mannoch Road will not see the sun again until the trees are gone. I cross the upper reaches of the Rothes Burn and suddenly the track is transformed into a sunlit forest highway that pushes back the surrounding woodland. This is the service road to the new Rothes wind farm. I count thirty-nine white turbines standing tall on the nearby hilltops, eating up the wind blowing out of the north. I like to reimagine the past and it would have taken the distillery workers with their horse-drawn carts two hours to walk here from Cardhu, to reach a point where they could take a short break and contemplate the day's work ahead. Below them was the Birnie Moss, one of the largest peat-cutting areas in the Speyside region.

At 300m above sea level, peat forms here primarily because of the high rainfall and poor drainage that creates the right level of bogginess for its growth. The view to the north would have been familiar to the distillery workers – rich agricultural land with the track descending through the villages of Shougle and Birnie to Glenlossie Distillery. Beyond, the town of Elgin, the destination of the Cardow Farm's illicit whisky. The Covesea lighthouse on the Moray Coast, today picked out in a patch of bright sunlight, would have been a familiar landmark. On a clear day, the distant view would stretch all the way up the coast to Brora and sometimes beyond to the herring fishing town of Wick in Caithness.

Although the arrival of coal in Elgin in 1754 via the seaports of the Moray Coast[48] led to a decline in peat use, Birnie Moss continued to provide an important fuel source for the local community and its distilleries for 250 years or more. At the end of the 1700s, peat was carted twice a week down the Mannoch Peat Road to Elgin from the Moss of Birnie and from Dallas, further up the River Lossie, to be sold in Elgin for eight to ten pence a cartload in summer, and up to fourteen pence in the winter.[49] At one time, Cardhu, Glenlossie and Longmorn distilleries were all using Birnie Moss peat before its use declined in the twentieth century and the old road eventually fell into disuse.

Up close, the Moss of Birnie begins to tell its own story. This was a big operation and the bog was drained in a planned and systematic way. The main peat-cutters' road meets three raised and parallel side branches that stretch out into the bog, and at one time would have carried a light railway to facilitate extraction of the peat. Little bare peat now remains and the bog is steadily rewetting with encouraging signs of abundant new *Sphagnum* growth. Heather and lichens colonise the dry areas and Sitka spruce seedlings are slowly moving into the area from the surrounding forest. Today, the moss is full of bird life; skylark, red grouse, golden plover and willow warblers all compete to make themselves heard. Unlike many old mosses I have visited, tree stumps and ancient bog wood are rare, a small copse of ancient pine trees growing here and there on drier small island hummocks surrounded by an open sea of wet *Sphagnum*-forming bog. On the surface, the signs of ecosystem recovery of the Birnie Moss are good, but the plants and birds here have had much longer to recover compared to the recently abandoned moss of Faemussach, just twenty miles away.

As the peat cutters of the Birnie Moss pushed further into the bog and extended their reach with their network of solid, raised tracks, they began to win the best and deepest peat. At the back of the bog, the peat reaches three to four metres in thickness, and at one time would have been deeper still. I find bare patches of soft and crumbly peat littered with small fragments of bleached wood. A few relics of the past are scattered around – a peat wagon, a discarded, deflated rubber Henley tyre and a well-made wooden bridge over a still active open drain. This was the last area of the bog to be worked and in 1967, Longmorn Distillery stopped using local peat and started using a new source. Under the heading 'Peat', the minutes from the September 1967 board meeting of Longmorn-Glenlivet Distillers Ltd record the moment precisely:

The deliveries of peat from the Northern Peat & Moss Co., New Pitsligo, have been most satisfactory and since no other distilleries are now using the Mannoch Hill Peat Moss the cost of maintaining the roads and bridges thereto could be prohibitive. It was decided to vacate the Mannoch Hill Moss and to advise the superiors accordingly.[50]

Speyside peat had been replaced by Aberdeenshire peat and the Mannoch Peat Road was abandoned. Three years later, Longmorn Distillery operated its floor maltings for the last time.

But the influence of the Mannoch Hills and the Birnie Moss lives on in its water. Apart from supplying Cardhu Distillery to the south, a brown peaty stream rises in the Mannoch Hills called the Gedloch Burn, an important water source for both the Mannochmore Distillery and its older co-joined sister Glenlossie. Mannochmore was only built in 1971 and produced the notorious Loch Dhu, the first black whisky, a blackness that many suspect was produced not by excessive use of peaty water but by copious amounts of added E150 caramel.[27] It has developed a strange cult following, despite being variously described as 'horrid', 'nauseating', 'a gothic horror' and best of all, *'aqua crematoria'*.[51] Its 'success' encouraged a Danish independent bottler to create *Cú Dhub* (Black Dog in Gaelic), a whisky produced by the Speyside Distillery and blackened with E150 caramel, to make it the 'only black single malt whisky in the world'. Its only award was a Gold Medal in 2013 at a whisky fair in Finland.

The quiet solitude of the day is broken by a pair of Eurofighters lining up to land in formation at Lossiemouth air base on the coast. I retrace my steps and leave the Birnie Moss to continue to regrow and rewild and head back to the upper reaches of the Rothes Burn, another peat-rich stream rising in the Mannoch Hills. As I follow the burn downhill towards its confluence with the River Spey, I pass by six-metre-thick peat banks and walk along well-made tracks through huge areas of forestry, some recently felled, some in the process of being felled by noisy diesel-powered harvesters. The disturbance in the catchment all adds to the water colour, and as I finally reach the flatland of the Spey valley and walk past the stillhouse of the Glenrothes Distillery, the burn is a fabulous deep beery hue. When he visited in the mid-1880s, Alfred Barnard simply noted at Glenrothes that the source water from the Mannoch Hills had a 'brown tinge so common in the Highland stream'.[17] A rare understatement from the chronicler of distilleries: this is one of the most coloured, peat-rich burns I have ever seen.

Glenrothes Distillery had its first spirit run on Sunday 28th December 1879,[52] the same night the Tay Bridge collapsed in the great storm that claimed the lives of all seventy-five passengers and crew on the Wormit to Dundee train. It was the same storm that flattened mature forests across Scotland, ripped the roofs off buildings, toppled castle towers and monuments and, when it arrived in Speyside, blew down the chimney stack at Balmenach Distillery and sent it straight through the roof of the stillhouse.[53] During the day, lighthouse keepers on the coast had recorded dramatic fluctuations in temperature and pressure as the eye of the storm passed across Scotland. At 9pm that night, after wreaking havoc on the Highlands, it headed off across the North Sea towards the 'peat islands' – the Norse flatlands of Orkney and Shetland.[54]

The Peat Island

'Where are you flying to this morning?' asked the Loganair rep at Kirkwall Airport.

'Eday,' I replied, simultaneously balancing my backpack on the baggage scales and handing over my boarding pass. A blank stare instantly told me all was not right. After several flourishes of the keyboard, each followed by a short pause, he announced, 'There are no flights to Eday today. They don't have a crew to man the fire engine. Would you like me to book you on the next flight?'

'When does it leave?' I asked. Another flourish of the keyboard and then his answer: 'In ten days, when the fire crew are back'.

I declined the invitation, retrieved the backpack and binned the worthless boarding pass. My plan to land on the dirt track of Eday's London Airport on midsummer's day, dashed. I asked if there was a ferry to Eday today.

'Yes. There is one at 7am,' but after checking his watch he announced, 'You've just missed it'. Temporarily grounded, I called the ferry terminal at Kirkwall and booked a ticket on next morning's crossing.

I spent the evening ruminating in Stromness hoping that tomorrow's attempt to reach Eday would not meet a similar fate. Walking along the cobbled streets and quayside wharfs of the old town that resonate with memories of the Hudson Bay Company and Arctic whaling, I was rewarded with one of those special evenings that only midsummer on a northern island can serve up. Brilliant light, still seas, soft distant voices and laughter – I could have been in Reykjavik, Tórshaven ... or even Eday.

The following morning, I stepped off the MV *Varagen* at Backaland pier in bright sunshine on the south end of Eday, the only foot passenger to leave the gaping mouth of the ferry. It swallowed up a party of excited school children and was off again to neighbouring Sanday. Parents waved goodbye and the hubbub of inter-island arrival and departure quickly melted away. I was left alone – becalmed and happy in a peaceful, beautiful place. I had finally arrived on Eday, a 12km-long (7.5 mile) island in the northern Orkney group, once home to a major maritime peat trading operation that supplied the peats that fired the kilns of many of Scotland's most famous whisky distilleries for 100 years.[55] From the sixteenth century until the 1940s, the islanders traded their famous peats, the 'inkies' and 'yarphies', firstly to local islands and then, as demand and reputation grew, across the seas to the distilleries of mainland Scotland and maybe even as far away as Australia. I wanted to walk the quiet roads and footpaths of this long, thin, 'once famous peat-exporting island' that had quietly slipped out of the limelight to see for myself what remains of its rich peatland past and write its story.

I walk north across the narrow sandy low-lying connection that is the home to the island's deserted London Airport and its unmanned fire engine waiting patiently for its returning crew. Clusters of farmsteads come and go, many abandoned and roofless, some working. At one time, each of these farms would have had a peat-fired stove and an open peat fire. Many had a peat kiln in the yard to dry the damp grain, which in some years would not be harvested until late November. Shredded peat and powdery dross would be used as winter bedding for cattle and sheep, and each day children from the farms would carry a single peat turf to the school classroom to help dry their clothes and keep their bare feet warm.

Peat formation on Eday is thought to have started 3,000 years ago; the island is one of the few places in the British Isles where the arrival of humans in Neolithic times predated the arrival of peat. These first peoples left behind lichen-covered standing stones, burial sites that over time were plundered for their treasures, chambered cairns, waste-filled middens and beneath the younger layers of peat, buried Neolithic stone dykes.[56] After the trees were cut down, peat from the higher ground became the primary source of fuel for Bronze and Iron Age settlers attracted by the island's low-lying fertile land. These early peoples were followed by Picts, Vikings and latter-day pirates. By the 1600s, Eday had an important kelp industry, sandstone quarrying, crab and lobster fisheries, and a seventeenth-century salt industry that burnt large amounts of local peat to evaporate seawater in open pans. At one time, its population numbered 1,500, but by the end of the 1800s it had declined to 650 – the most recent census returned a permanent community numbering just 150.

On the island's west coast, I pick up a high cliff path and head north towards Red Head. Far out on the other side of the Sound are the islands of Westray and Papa Westray; next landfall some 200 nautical miles across open ocean is the peat-rich archipelago of the Faroe Islands. The views from Red Head are simply huge and I stumble upon the concrete foundations of a World War I submarine lookout hut. Not until I round Red Head on the northern tip of the island and begin to walk south along Calf Sound do

I begin to see the regular shapes of the old workings. On the east side of the island, the peat banks are dry and concealed under a carpet of heather. Many stretch out for almost a quarter of a mile – the oldest and some of the longest sets of cuttings on the island.

Eday's maritime peat trade can be traced back to the late sixteenth century when 'peitts' from Eday were exchanged for corn grown on neighbouring Sanday,[57] where peat had become so scarce that the locals had to burn cow and horse dung to supplement their meagre fuel supplies. Eday quickly realised that it had its own 'black gold' and in the 1630s set its own gold standard, the Eday Peat Fathom. Measuring twelve feet long, six feet wide and six feet deep, it was equivalent to a peat volume of $12.3m^3$ or a dry weight of just under four tonnes. Traditionally, the individual islands would have had an annual peat-cutting day, but by the late 1700s, many had begun to run out and they began importing and buying fuel from peat-rich Eday. Coal started arriving in Orkney in the late 1700s, but it could only be afforded by the more well-off families in Kirkwall.

Eday's peat trade was now in full swing, and in 1828 the estate and its workers cut and sold 1,742 tonnes dry weight for export. Eday barley was already being shipped to the Firth of Forth, and in the 1840s the start of a regular streamer service to the Port of Leith allowed Eday peat to travel south to Scotland's capital city, where it was bought and sold by quayside merchants. Some of it would have found its way to distilleries in the Central Belt, but it was not until the 1880s that records began to show direct sales to named distillers.

In the early 1850s, Eday had a new laird and a change of ownership that was to prove significant in the distillery peat trade. Bought for a knockdown price of £9,000 at an Edinburgh auction from a debt-ridden owner, the island came into the hands of Robert James Hebden, a wealthy landowner from the south of England. The sale notice highlighted the profits to be made from the island's peat resources on 'which the neighbouring islands depend for fuel, and from which peats are exported to other parts of Scotland'.[58] Successive generations of the Hebden family have been the sitting lairds and main landowners ever since.

The inter-island peat trade reached its peak soon after the new owner arrived, but the wider availability and lower price of coal in the 1880s brought a decline in local demand. This coincided with a rise in trade with the whisky industry, and in 1881 the estate's peat accounts show a net income of £120 with shipments to distilleries in Liverpool, Aberdeen and Fife, the largest of all to the Adelphi Distillery in Glasgow. By 1891, 1,320 tonnes of peat left Eday destined for ten distilleries, shipped south on a new fleet of large steamships. The biggest customer was the DCL and Eday peat was now finding its way to a swathe of distilleries across Scotland – Cameron Bridge, Glenkinchie, Ben Wyvis, Mortlach, Highland Park, Strathisla, to name just a few. The years 1896–8 were a boom time for Eday's peat trade with a healthy annual profit of £425 and almost 2,000 tonnes of peat supplied to twenty-one distilleries, the largest shipments of peat ever to leave the island. Orkney peat, and most likely Eday peat, crossed the world and in 1893 reached distilleries in Australia through the Port of Melbourne.

In the last ten years of the nineteenth century, over thirty distilleries are named in the estate records as purchasers of Eday peat, including Ben Nevis, Glenmorangie, Pulteney, Glenlivet, and the three Inverness distilleries of Glen Albyn, Glen Mhor and Millburn. The arrival of World War I put a temporary halt to the local peat industry as men went to war. Local men, women and children now worked the bogs in a different way, gathering *Sphagnum* moss that was packaged up and sent south to make wound dressings for soldiers in the European theatres of war.

After the end of the World War I, labour on Eday was scarce and Major Harry Hamilton Hebden, the returning laird and grandson of the original owner, began to break up his estate. He talked to his great friend Walter Grant, the owner of nearby Highland Park Distillery, about the future of Eday's peat and made plans to restart the distillery peat trade. The Eday Peat Company was formed in 1925, 'carrying on the business of Peat Exporters', with the aim of commercialising the operation and attracting peat cutters to Eday with the reward of better pay and working conditions. Harry formed a business partnership with local laird and fellow soldier Colonel Henry W (Billy) Scarth, and responding to the competitive nature of the post-war peat market, they sent samples south to a laboratory in England for chemical analysis. The results showed that Eday peats were of high calorific value, unusually rich in carbon and low in mineral impurities such as nitrogen and sulphur. In short, they were excellent peats. Letters to distilleries quickly followed promoting the now certified quality of Eday peat as the new company sought to re-establish its pre-war markets.

The Eday Peat Company swung into operation in early 1926 when a steamship arrived carrying a deconstructed army surplus hut, to be reassembled and provide a base for seasonal peat workers. A sandstone jetty was built near the Calf Sound lighthouse, and the peat hut equipped with the precision of a military operation. Two sets of rails ran down to the sea and the trucks, one full, one empty, operated a push-me, pull-me system up and down the jetty. The peats were tipped into holding nets on flat-bottomed boats that were rowed out and then lifted onto waiting steamships in the sound.

Things started well for the newly formed company with an order for 1,100 tonnes of peat from the Dalmore Distillery. Ten men were employed on the peats for four months a year, working forty-eight hours a week. The important customers quickly returned — The Glenmorangie Distillery Company, Mackinlays & Birnie Ltd (owner of Glen Albyn and Glen Mhor) — and large shipments of Eday peat began landing again on the beach at Glenmorangie and at the wharfs of the canal basin in Inverness. In 1929, Scottish Malt Distillers (SMD) received 750 tonnes, the largest single order that year.[59]

Profits in the early years were good, but in the years leading up to World War II, demand began to decline, profits plummeted and the peat fields on Red Head became all but exhausted. *The Scotsman* newspaper was already writing the company's obituary,[60] calling it 'the once famous peat-exploring island'. The Eday Peat Company accounts for 1937 show that only 300 tonnes of peat were cut, returning an annual profit of just £66. The company was clearly struggling and customers were becoming unhappy with the quality of the peats. James Birnie, a Director of Mackinlays & Birnie Ltd, wrote[55]:

> We took delivery of the three cargoes of Peats totalling 185 tons … but we are very disappointed with the last two in which there was a large amount of dross, as well as fibrey Peat. We have asked you before to see that we got the hard, black Peat and in future you must send us not so much dross.

World War II brought an upturn in the company's fortunes as scarcity of coal drove up demand for local peat. It supplied peat to military units and several distilleries, including Scapa and Highland Park, became good customers. The price of Eday peat was now 29s a ton, double what it was fifty years earlier. But as the focus of peat production and shipping moved from the north end of the island to the mosses and larger piers in the south, the company was again in trouble, barely operating at a profit.

As I walk past the old heather-covered abandoned peat banks on Red Head on a warm afternoon, it is hard to imagine this place in the frenzy of a summer's peat harvest. Most distilleries wanted the top two feet of yarphies, rough peats, while the remaining peat down to a depth of six feet, the inkies, dense and hot burning, was destined for the home fire or the school classroom.

After descending the cliff path, I walk inland following a wide, old track. In a wet depression, I find open peat banks, probably worked last year or the year before, their tops protected by a line of brittle bleached wooden pallets that crack and splinter under my weight. The banks are 1.5m high, light-brown, dry, cracked and fibrous on top, full of rootlets and plant stems; deep down dark-black, wet and amorphous – Eday's yarphies and inkies. Retracing my steps along the old peat cart track, I soon arrive at a pebble beach by the Calf Sound lighthouse, and sticking out from the grassy hummocks and exposed by the sea are rusting iron tracks and the shell of a tipping truck, all that remains of the long-abandoned peat railway. As the tide ebbs, the seaweed-covered skeleton of the sandstone peat pier slowly reveals itself, where two sets of rails once carried the peats the remaining yards to waiting boats. It fell into disuse in the inter-war years after suffering storm damage. I decide to camp here for the night amongst the relics of the old peat railway.

All evening fulmars glide back and forth along my coastal dune. Clouds begin rolling in from the west signalling a change in the weather. As I eat what I could find in my rucksack, I become aware of a roaring sound like an approaching train. The tide has begun rushing through the narrow gap that separates Eday from Calf Island, generating whirlpools and standing waves. The locals call this the Eday River, a rip tide generated by huge masses of water that ebb and flow between the Atlantic and the North Sea. Peat ships would time their departure from the sound using the tides to catapult them on the first leg of their journey south. The islanders once had a chain stretched across here and when the roar had subsided, they used it to drag a barge full of peats across Calf Sound to the main island.

Later that evening as I pulled out my hip flask, I was reminded that for more than 100 years, Eday was a dry island. In the 1850s, the new owner had closed the alehouses, 'no bothy or whisky shop is allowed there', and the island became a place 'where the sale of stimulating beverages is abolished by general consent'.[61] It also became a place where people were sent from neighbouring islands for 'the cure'.

In the early hours of the morning, I awoke to misty rain and poor visibility – the kind of Orkney midsummer weather I am used to. I packed my tent and walked south past a silent and unlit Carrick House, still the family home of the Hebdens. Major Harry died here in September 1939. He had been wounded and badly gassed at the end of World War I, and like so many soldiers who served in the battlefields of northern France, never fully recovered. On leave in Eday just before the onset of World War II, the laird of Eday caught pneumonia whilst out sailing and died in the house where he was born.

Major Harry was just fifty-three and his passing signalled the beginning of the end for the fledgling Eday Peat Company. After the end of the war, as distilleries accelerated production, Glenmorangie and Strathisla returned with more requests for peat, but in the climate of post-war Britain, Major Harry's widow struggled to restart the business. A proposal from a Glasgow businessman to start a mechanised operation came to nothing, no peats were traded in the 1950s, and the company was finally wound up in 1965 with assets valued at £88 4s 6d that were split between the two shareholders. And so ended the remarkable story of a short-lived maritime peat-trading company that during its heyday influenced the taste of whisky made at one third of Scotland's operational distilleries. Exploiting the well-established shipping lanes, Eday's peat found its way to the east-coast distilleries, to the west coast, inland to the Highland distilleries of Perthshire and Speyside, and south to the Central Belt and the big cities.

The rain had begun to clear as I return to Backaland pier at the south end of the island. Shuttling back and forth across the north islands, the MV *Varagen* returns out of the mist from neighbouring Stronsay and we're soon heading back to Kirkwall, once home to the offices of the Eday Peat Company. In the distance I make out the low contours of Sanday, Stronsay, Shapinsay and mainland Orkney – all former destinations for Eday peat. We are following the first part of the route once travelled by those peat-carrying steamships of the past – the SS *Millrock*, SS *Jesmond* and SS *Orcadia*.

So what made Eday peat so special? Maybe the last word should go to Jimmy Garson who has been working the mosses for maybe sixty years; one of very few folk still cutting peat on the island. Asked by a reporter why Eday peat was so attractive to the distilleries down south, he replied thoughtfully, 'I really do not ken that, but they were very good quality peats, all right.'

Three

Peat Shed to Electric City

'Welcome to California, Scotland – The Sunshine Village'
The sign may have stopped you in your tracks but if there was still
any doubt as to which country you are in, the village's welcome
includes a large unmistakably Scottish green and purple thistle.
When the sign first appeared, designed by the local council in
a humorous moment, it contained a bright yellow Californian
sun. The topcoat of paint has been removed by time and all
that remains is the undercoat, a dull, milky white Scottish sun,
uncertain whether to rain or shine. At the local football field the
faded words 'Welcome 2 Hell', white-painted on the wall of the
blue Portakabin changing rooms, were less welcoming to visiting
teams. A single set of rusting goal posts suggests the opposition
was last subjected to this version of hell many seasons ago.

Nobody is sure how this former coal-mining village got its
name, but its rows of four in a block miners' houses, much loved
by Scotland's planners in the 1930s, tell of a hard-working community. Each blockhouse was home to four families with two separate
entrance doors at either end and eight windows front and back. A
mineral railway was built in the nineteenth century linking the local
coal mines and the local towns to the Union Canal with its distillery
wharfs. Employment could also be found in mining peat at nearby
Gardrum Moss, which was still an active extraction site operating
into the early part of the twenty-first century. It had its own narrow-gauge peat railway and like many similar bogs, spawned local
industries such as peat moss litter and drying-compression works.

In the mid-1880s on his tour of the whisky distilleries of
the United Kingdom, Alfred Barnard arrived in the thriving,

industrialised Central Belt of Scotland. He visited St Magdalene Distillery in Linlithgow and wrote[1] that its peat was 'obtained from the moors above Falkirk and round Slamannan'. Its peat sheds were filled from canal barges arriving at the distillery, and peat was combined with a precise amount of locally produced coke and then used to fire its two malt kilns. St Magdalene started making whisky in 1798, and after a long period of almost continuous production became, at the start of World War I, one of the five founding members of SMD. At times, it was quite a smoky, peated dram, different from the typical softer Lowland style of whisky. In the 1960s, it carried out experiments on peating with SMD's laboratory at nearby Glenochil.[2] St Magdalene operated its own floor maltings until 1968, after which it began sourcing its malted barley from Glenesk Maltings on Scotland's east coast. Like many other distilleries at the time, it fell under the SMD hammer and closed for good in 1983.

St Magdalene was a big, progressive, energy-hungry operation that required large quantities of fuel. In the seventeenth century, Linlithgow was already a major centre for milling and brewing, and at one time there were five distilleries operating in town, all sustained by local water, barley, coal and peat.[3] In 1834, the distillery was moved to the banks of the newly completed Edinburgh–Glasgow Union Canal and had its own dock. This was a clever piece of business planning. Along with its proximity to road and rail links, it meant that the distillery, which produced the 'famous St Magdalene Spirit', was now perfectly positioned to exploit the growing markets in the big cities. Raw materials also became cheaper and more accessible via the new canal systems. In 1870, the local newspaper ran a sketch of the distillery:[4]

The peats used in the distillery are obtained from the neighbourhood, mainly from the direction of Slamannan, and are found to be as good as any to be obtained. Certainly, the flavour imparted to the produce of the distillery is neither harsh nor so offensive as the peat flavour occasionally is, and so far, the quality of the peat used appears to be satisfactory.

A For Sale note in 1874, after the death of one of the owners, highlights its proximity to the Union Canal, which 'brings coal and peats at cheap rates'.[5] Both the Union and Forth & Clyde canals were important arteries for whisky producers in central Lowland Scotland, transporting barley, coal, peat and casks to and from the distilleries. Now silent, Port Dundas, Dundashill, Littlemill and Lochrin were all located at the end of the canals, or close to the canal basins in city centres. Along the canal routes lay St Magdalene, Rosebank and Bankier distilleries.

Peat obtained 'from the moors above Falkirk and round Slamannan'? This was a peatland area of Scotland that I knew little about, but one that had clearly played a significant role in the whisky produced by some of the great, now silent, Lowland distilleries. I set out to see what remains and head west out of Edinburgh past the huge landmark rusty brown bings – all that is left of 15 million tonnes of spent oil shale waste. After exiting the M9 when the unmistakable aroma of the giant Avondale landfill site wafts through the car window, I travel south, cross the Edinburgh–Glasgow mainline railway, then the Union Canal and head uphill towards California. As I get higher, the panoramic views of Grangemouth to the north, Scotland's largest industrial complex based on oil and gas and their refined products, gets better and better. Just across the water from Grangemouth is the giant, soon to be demolished chimney of Longannet, Scotland's last operational coal-fired power station. I am now just a few miles south of the Antonine Wall, a 63km-long (40-mile) turf dike built by Emperor Antoninus Pius as a defensive line against the troublesome Caledonian tribes around 142AD. It marks the northern extremity of the Roman Empire, plugging the narrow gap between Scotland's east and west coasts.

Leaving California behind, I walk west out of the former pit town on an old peat-miners' path to explore, in Barnard's words, the 'the moors above Falkirk and round Slamannan'. During the day, I will walk 24km (15 miles) along miners' paths, quiet roads, farm tracks and across rough unmarked land before arriving late in the evening at the 1950s sprawling, modernist new town of Cumbernauld. As I walk quickly past and beyond football 'hell', the land begins to get wetter and I enter the peatlands. I pass close to a huge pair of wind turbines compulsively and rhythmically chopping the air as they harvest energy in the early morning breeze. Up close, they sound like two small boats heading up into the wind, each one hitting wave after wave in quick metronomic succession. Now high up on the moors, I can see wind turbines in all directions, towards Glasgow to the west, far away to the north across the Forth estuary, to the east towards Scotland's capital and in the hinterland to the south. A moving 360-degree array of fast- and slow-spinning blades. It makes sense to put them here, close to places where humans live and work, where energy transmission costs are low, and on reclaimed, overused and often exhausted land. I like wind turbines a lot, and I like what they do. Alone, as identical twins, triplets or clustered in farms, they can bring visual energy to bleakness and enhance natural beauty.

I am surprised by the abundance and variety of wildlife I encounter in this wet, post-industrial landscape. A silent group of three roe deer cross the path in front of me, carefully picking their way through young trees, unaware of my presence. Maybe we humans don't pass this way too often. Chaffinch, goldfinch and coal tit advertise their presence loudly. A willow warbler keeps pace with me for a while, flitting from tree to tree in the scrub next to my path on the old peat railway. Then the unmistakable flash of a sparrowhawk, a top predator hunting low down and under cover in the flooded drains of the old peat workings. Colonisers are leading a charge of natural rewilding; birch, willow, broom, gorse and rhododendron, a pair of invaders standing isolated and alone on bare horticultural peat only recently destined for the garden centre. Although no attempt has been made to speed recovery here, grasses, sedges and mosses are beginning the slow process of taking over the brown, bare surface of the peat — a natural greening of the bog. The upturned relic of a rusty wheelbarrow with only the front wheel and its two broken legs still visible through a thick carpet of young moss tells that the bog will reclaim, given peace and time.

This is Gardrum Moss, most recently operated by Sinclair Horticulture and still actively extracting peat in 2008. Although much deeper in places, it had an average peat thickness of four metres and at one time contained 6.5 million cubic metres of peat.[6] Old workings are visible all around, a series of equally spaced parallel lines intersected at right angles by another set of neatly aligned cut marks; lines of drainage ditches, peat banks, tracks or the imprint of a narrow-gauge railway. At Gardrum Moss, machines created a framework of oblong extraction cells, each one separated by water-filled ditches that flood in winter and dry in summer. A natural non-linear landscape transformed into a linear man-made construct.

I come across another graveyard of ancient pine trees; their remains contorted, entangled and bleached. Roots, trunks and stumps unceremoniously dragged from their acidic, oxygen-starved, cold, wet resting places deep in the bog by men and machine, gathered together in a mangled mess to finally rot in the warm, sunlit, oxygen-rich atmosphere. These cadaverous relics, unearthed and disinterred by generations of peat miners, were an imperfection that had to be removed and cast aside. To others, they were a welcome source of light, the still resinous pinewood burning brightly when ignited.

When I visited in spring 2019, Gardrum Moss was still for sale. The entrance to the site was marked by a faded, barely legible red sign naming the former operators and a larger and brighter blue and yellow sign advertising the opportunity to buy 160ha of 'Agricultural/Amenity Land'. It is not easy to journey through these hills and moorlands high above the densely populated Central Belt of Scotland. Public roads come to an abrupt halt barred by locked gates and deliberately bulldozed banks of rubble...Keep Out, Go Away, Do Not Come Back...restricted access, barriers, fences, barbed wire, razor wire, CCTV in operation. This speaks to me of control, past ownership and wealth, hazards, fear and danger, a mess waiting to be cleaned up, an uncertain and unplanned future. No welcome here to the Sunshine State. The area is still a mecca for illegal fly-tipping of commercial and household wastes. Tyres and burnt out vehicles are not an uncommon sight. I sense the land is tired, worn out, scarred, unloved and uncared for, forgotten, its surface broken and punctured, the wounds unattended with no time to heal. Humans have come, taken away, left, returning occasionally under the cover of darkness to soil. Scar Hill, Thieves Hill and Jaw Hill maybe tell of an even darker past. A land in need of love.

As I head west across the hill country with its sparse livestock farms and soggy fields, picking up a track along an occasional dry ridge of sandstone or limestone, the peatlands begin to arrive one after another. Gardrum Moss is followed by Darnrig Moss, Garbethill Moss and finally Fannyside Moss. I walk through an old farmyard that had morphed into a hybrid place of second-hand shipping containers, a recently occupied caravan, a discarded pink child's bike. There is no one at home, apart from two fat and healthy-looking ducks in a large cage.

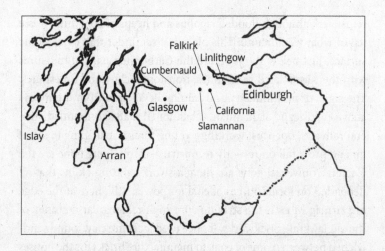

It's a little harder to see past evidence of peat extraction at Darnrig Moss but look closely and the signs are there. A series of parallel ridges, like the ripples of a corrugated iron roof, each ridge exactly twelve metres apart and now completely carpeted in thick heather moor. Peat with a thickness of almost ten metres was recorded here when the bog was last surveyed.[6] Deep hazardous drains, still active and serving the purpose for which they were constructed, cut the ridges at right angles, drying the peat and creating a perfect place for heather to grow. The familiar stop-start, crackling call of a single red grouse confirms the transformation and destiny of the land, and now in bright sunlight, I meet a fox locked on to its prey. I see it before it sees me. A quick look up and it dissolves without sound into deep heather. Maybe we humans don't pass this way too often.

Darnrig Moss has had a long time to recover since it was an active extraction site supplying peats to the distilleries of Linlithgow more than 100 years ago. However, in the 1980s, the eastern part of the moss was ripped open again, when for a short time it became an opencast coal mine. It has gone now and the land has been reinstated, although flooded lagoons and heaps of coal mine waste tell of what went before. This place is a reminder that close to the surface, just below the peat, are the Carboniferous coal measures with their associated sandstones, ironstones, clays and limestones. The signs of past industry and mining lie all around in the moors above Falkirk and Slamannan – abandoned shafts, dismantled mineral railways, opencast workings, young forests struggling to stand up and grow tall on recently reconstructed soil. Long gone are the Carron ironworks, gone are the glassworks and limekilns that all depended on local sources of coal for power. Still, here at the edge of Darnrig Moss is the striking 30m-high rectangular chimney of the old Jawcraig brickworks. It once used local fire clay, found sandwiched between seams of coal, to manufacture bricks for the houses of coal miners. The works closed in 1998 – the old buildings now form part of an organic waste storage and treatment operation.

Downhill from Jawcraig and just outside Slamannan – the name perhaps derived from *Mannan*, the Celtic Bog God – I cross the River Avon, now a restored, free-flowing natural waterway; a reminder that surface water played an essential role in this industrial landscape. Some still manages to follow its natural course, but much has been diverted, trapped and controlled. Drains, culverts, canals and dams rerouted the water as impoundments, ponds and pools retained and trapped the water. A source of flowing water was vital to the local industries: a method of transportation, a source of power, a conduit for waste products and waste heat, a coolant, a liquid for industrial chemistry and processing. Whisky making needed water to drive waterwheels, to steep and mash barley, cool the distillate in condensers and worm tubs, reduce the strength of the spirit, wash and clean, and in time to make steam.

Located at the heart of this coal-rich area and close to the Jawcraig colliery, the town of Slamannan was once famous for its highly skilled miners. It had its own railway station, a saddlery, bakery, miners' co-operative, brass and pipe bands, Masonic hall, five pubs and a hotel. Across the street from the Boer War memorial clock is the St Laurence bar, the only remaining pub in town. In search of a pint to rehydrate I go in and, being the first customer of the day, am made welcome by Lorraine, the manager. We talk about the village, its mines, the past and the present: 'Everybody worked in the mines, my faither and my faither's faither. Many of them worked in the wee mines.'

'There were also peat miners,' said Billy, a retired coal miner and Slamannan man, who had dropped in for his lunchtime pint. 'Fannyside Moss was big and it was the Dutch who came over – they owned and ran the operation. For a day in the summer, the folk from Slamannan would put on their best clothes and walk up the road to Fannyside Loch for a picnic. That was our holiday for the year.' Billy chuckles when Lorraine reappears with a copy of an old newspaper article with the heading 'Miners wanted – Slamannan men preferred'. Local folklore has it that an advert bearing the same title was nailed up outside the new mining camps springing up in the Canadian Rockies. Slamannan miners working in the 'wee pits up the braes' were expert 'handpicksmen' and highly skilled at winning precious coal from thin and complex seams. Many left for North America, escaping the poor working conditions and pay that led to the Slamannan Riots in 1878. The average life expectancy of a miner here in the mid-1800s was thirty years.

It is time to leave and Lorraine refills my water bottle. I thank her and Billy for their hospitality. 'We'll be seeing you again soon, and don't forget to phone if need be.' After telling them I am walking to Cumbernauld via Fannyside Moss and explaining why, Lorraine gives me both a sideways look and a phone number to call in case I got into trouble. I leave Slamannan, and back on the high road continue west along the edge of Garbethill Moss. Parallel cut marks across a peat-rich landscape softened by a thick blanket of heather tell a now familiar story. For a while, I am accompanied by the crack of rifle shot and the thud of shotgun fire and exploding clay pigeons. Garbethill Moss has become the home of the National Shooting Centre. I walk quickly on. Tortoiseshell butterflies are up and about and as the afternoon becomes warmer, I arrive at the eastern edge of Fannyside Moss, the largest and most impressive of the moors above Falkirk and round Slamannan.

More than nine metres thick and covering over 500ha, peat started accumulating here more than 9,000 years ago. The bog is almost perfectly flat, making it an ideal site for industrial-scale peat harvesting and a great place to get lost. At one time, it was worked by the Fannyside Muir Peat Extraction Company, who used strip-mining methods to harvest the peat, layer upon layer. A satellite image shows the most recent surface mining activity: a striking symmetrical pattern of equal sized 10x40m extraction cells, each one separated by a drainage ditch. These smaller identical cells make up the constituent parts of much larger extraction blocks measuring up to half a kilometre wide. Standing in the bog, it is impossible to appreciate the scale and near-perfect symmetry of this man-made landscape. Unsurprisingly, walking through the moss is foolhardy and I regret it immediately – a challenging assault course of ditches and drains. Some are three metres deep and four wide and more than 100 years old. Thankfully, I reach the old peat railway line, raised above the bog on wooden sleepers. The rails and point systems are still just visible, but many are pushed aside and buckled by growing willow scrub. The displaced rails are fast disappearing beneath bushes of flowering yellow gorse. A new layer of *Sphagnum* moss, soft and thick like a deep woollen carpet, covers the damaged peat below.

After commercial peat extraction at Fannyside Moss ceased in 2013, the processing factory and its peat wagons became a favourite play area for local children. It also achieved some notoriety as a popular venue for illegal acid house parties. Bog repair work began in 2014 and, under the guidance of Buglife Scotland, more than 210ha of degraded peat moss have been deconstructed and restored by volunteers working with local contractors. While the theory of bog restoration is relatively straightforward, each bog is different and has its own challenges. Drains and ditches are blocked, dammed and re-profiled, invasive trees and scrub removed. This raises the water table, rewetting the bog and creating the right conditions for recolonisation by peat-forming *Sphagnum* moss. What this doesn't do is clean up the peat in an area that was once subject to high levels of atmospheric pollution from local industry. Beneath the surface of newly formed peat at Fannyside Moss will always be a long-term record of the acid rain, heavy metals and particles that once rained down from the atmosphere onto the bog from nearby power stations, refineries, metalworks and mills. The impact of the restoration work at Fannyside Moss and the recently constructed 3,700 dams is immediately impressive. The bog is wet again and the regrowth of *Sphagnum* so fast that many of the artificial dams are already invisible. There is also the unwanted evidence of the scars of a recent heather and grass fire.

Like Duich Moss on Islay, this Lowland raised bog has been afforded a level of protection and restoration because of visiting geese, this time the Taiga bean goose rather than the Greenland white-fronted goose. Each winter, the shallow pools and lochans of Fannyside Muir become home to a regular group of over 200 of these welcome visitors from the Arctic, the largest single over-wintering population in the UK. Not quite the famous whisky versus geese battle on Islay of the 1980s, but a more recent skirmish between peat extractors and conservationists, the latter empowered by a judgement based on internationally binding legislation on wildlife protection.

A high, elevated road crosses Fannyside Muir connecting Slamannan to Cumbernauld in the valley below. It reminds me of the long, straight and bouncy airport road on Islay that seems to float across the open peatland between Port Ellen and Bowmore. Like on Islay, drivers wave to a stranger standing alone in the bog. I reach the road, walk unnoticed past a car with two lads locked into their mobile phones, and drop down to the loch, where it was said illicit whisky was once made:[7]

> In our visit to the loch, we heard a good few incidents about what was called the 'wee stell', from which the real peat-reek flavoured liquid was produced, and were shown the ruins and remains of a rude distillery level with the moss on the north bank of the loch, whose threshold, we were informed, no gauger's foot had ever crossed.

I walk around the loch and find its four-metre-high peat banks on the north shore and peer into the black waters, but unsurprisingly find no sign of illicit distillation. It's time to leave the mosses of the high ground and head downhill towards the sounds of Cumbernauld below and the noise of traffic. It's hot, I am tired and have a train to catch. Before leaving the loch, I pass a burnt-out area of trees. Much further downhill, I exchange greetings with a man and his dog on a golf course, the first fellow walkers I had encountered since leaving California eight hours ago. In a woodland clearing just outside town, four firemen are beating down a grass fire. The weather has been dry and it's the start of the Easter school holidays.

The moors above Falkirk and round Slamannan are different from many of the peatlands in Scotland and a long way from the idyllic view that people hold of our bogs and mosses. There is a natural tendency among writers to over-romanticise peatlands and create an image of a hard but worthy and healthy lifestyle built around the annual cycle of the peat harvest in the clean air of the Highlands and Islands of Scotland, or Ireland. Here in the Lowlands and from early times, organic carbon in solid, liquid or gaseous form has been a source of wealth and power; extracted, scraped and squeezed from the land. The Romans, encamped here just south of the Antonine Wall, built roads and dykes, felled trees for fuel and made charcoal from the old forests. They were also some of the earliest humans to exploit the extensive peat deposits beneath the surface. This is a remarkable energy-rich place of Carboniferous coal, oil shale and natural gas that helped fuel Scotland's industrial wealth – wind turbines now harvest power from the atmosphere in a new energy revolution. The area is known to contain significant reserves of frack gas and coal, and maybe peat will be mined again. But this is a landscape that bears the scars of industry, an exceptional place rich in fossil fuels and natural resources, and about as far away from the misty-eyed view of a peatland that is possible.

As I cross the Forth and Clyde Canal on my way back to Edinburgh that evening, I am probably the only one on the train to notice the chimney of the old Rosebank Distillery. Unlike St Magdalene, which the train passes ten minutes later, this famous Lowland distillery is about to be reborn as I write this, more than twenty-five years after it was last operational. Rosebank, well known and much loved for its light, floral, elegant style of triple-distilled whisky, at one time would have been peated. Tasting notes of a 17-year-old, 1974 vintage Rosebank reveal a 'touch of smoke'.[8] Barnard tells us that its malted barley was dried with peat cut 'from a good moss ground within four miles' of the distillery and stored in a canal-side peat shed. Almost certainly, this early smoky Rosebank would have been made with peat cut by hand from the mosses of either Gardrum or Darnrig, through which I had journeyed earlier in the day. Later that evening, I read that the long gone Bankier Distillery located five kilometres to the north of Fannyside Moss had a 'huge peat shed, filled with peats dug in the Cumbernauld Moors'.[1] Each year in the late summer, cartloads of dried peats would have rolled north downhill along the Fannyside Muir road, passed through the small village of Cumbernauld, crossed the canal at Wyndford Lock and filled the huge peat shed. Bankier Distillery was supplied by canal barges filled with locally mined coal; it had a steam engine, two steam boilers with large belching chimneys and a 'gas house'. It was a product of the 1820s boom in distillery construction – built by coal and powered by coal.

Lady Victoria Colliery, Newtongrange, 1976

As we stepped out from the cage and boarded the underground train, it was like transiting through some subterranean London metro station from the past. Sixty seconds earlier, after the banksman had given us our check tokens at the pithead, the winding engine fired into life and dropped our stomachs at a speed of 20mph, plunging us down the shaft into the warm depths of the mine. We now found ourselves 1,600 feet below ground in a huge gallery of red-brick ceilings and high arches. Squeezing together on rough wooden seats, our bogies began to move silently towards one of several tunnels that disappeared into the darkness and vastness of an underground world.

The journey from pithead to coalface seemed to take an eternity as the roadway stretched further and further into the mine in search of the ever-receding coalface. I sensed we were slowly rising, silently following the incline of the Carboniferous strata in an upward trajectory. Our train was pulled by an 'endless-rope' haulage system powered by electricity generated by the mine's steam turbines and fuelled by coal dust from 'the Lady'. We switched our head torches on and off to examine the rocks, shifting uncomfortably on our hard seats as we travelled through the mine. And then, after a journey of four miles lasting one hour, the strange quietness was finally broken by distant sounds and vibrations that grew louder as we approached our destination, one of the six main coal seams.

In turn we left the bogies, bent our backs and moved up to the one-metre-high coal seam. Moving back and forth like some giant mechanised leaf-cutting insect, the 'continuous miner' chomped away at the coal face dropping the shiny black rock on to an armoured moving conveyor belt that swept the coal away to the waiting hutches. To protect the working miners at the coalface, hydraulic props followed men and machine forwards as the roof cracked and collapsed behind them; cramped, dangerous work in a dusty, hot and noisy place. This was exciting and intense, and I had never seen anything like it before or since.

More than an hour later, we stepped back out from the cage at the pithead into daylight, blinked, and returned our check tokens to the banksman. All accounted for, we headed for the miners' baths.

Scottish Power and Coal

Coal was the fuel that created steam and powered the Industrial Revolution. It started with a steam engine invented by Thomas Savery at the beginning of the eighteenth century, but it wasn't until 1776, when James Watt and Matthew Boulton significantly improved its design, that steam power triggered a revolution that swept through Scotland and across the world. In 1786, John Stein installed a steam engine at Kilbagie Distillery near Alloa and by the early 1800s, steam was powering factories, mills, mines and even airships. Kilbagie and its huge neighbour, Kennetpans Distillery, became symbols of Scotland's new industrial age and its dependency on coal power. The writing was on the wall for peat.

Steam power effectively unlocked the energy needed to drive large machines and in the 1880s, Glenmorangie became the first distillery to enter the steam age by wrapping its tall stills in superheated jackets that were filled by hot water vapour generated from coal-fired boilers.[9] Distilleries built before the age of steam needed the energy provided by water or wind to turn millstones and operate machinery; the force of gravity was often utilised and cleverly incorporated into distillery design (such as at Annandale Distillery) and many tasks were done manually. This meant that distilleries and mills were constrained in size, typically small and dependent for their kinetic energy on fast-flowing rivers or man-made weirs and waterways. The arrival of the steam engine brought to an end the dependence on wind, water and muscle power to turn windmills and waterwheels and push sailing ships across seas. Wagons and barges no longer had to be pulled along railways and waterways by humans or horses.

Native American tribes called the new steam-powered loco-motives 'fire engines', and the advent of steam changed the world. Thermal energy could now be transformed into kinetic energy and coal, that until then was only used to heat mashtuns and stills, could now generate steam to drive machines. Distilleries grew in size and a new generation was built in locations that didn't depend on waterpower, such as in the centre of cities. The massive distill-eries of Dublin City, like John's Lane and Bow Street, all spawned tall brick chimneys that vented noxious coal gases high up into the atmosphere. With the arrival of electricity in the nineteenth cen-tury, motion or kinetic energy could now be converted into heat. Energy could be sent 'by wire' from huge power stations powered by fossils fuels and the kinetic-thermal energy revolution was complete. Wood, charcoal and peat had given way to coal, then oil and finally gas. Towards the end of the twentieth century, electric-ity generation began to move away from reliance on fossil fuels to a new era of nuclear power and renewables, harnessing energy from the wind, sun and tides. Energy generation had become cleaner and the tall brick chimneys, like the pagodas of the malt kilns, became redundant. The future will be cleaner still.

PEAT SHED TO ELECTRIC CITY

Scotland is naturally blessed with vast resources of carbon-rich fuels that over time have allowed it to evolve and develop industries powered by an array of different energy sources. Wood, peat, coal, oil and gas have all been extensively and successfully exploited by a succession of inventors, technicians and industrialists, and much still remains below ground or at the surface. Scotland was at the forefront of the development of nuclear power and as we enter the post-fossil fuel, carbon-neutral age, it is well placed to generate its future energy needs from the power of its reservoirs, rivers, tides, winds and waves.

Significantly, the Industrial Revolution signalled the beginning of the end for the age of peat as a major source of fuel. Although peat had been used as a fuel to raise steam in herring boats[10] and to power railway locomotives in Massachusetts, Quebec, Bavaria and Oldenburg, on a power-to-volume basis it was never going to compete with coal. At the end of the nineteenth century, steam locomotives in Bavaria were using 15,000 tonnes of peat a year[11] and had to haul huge peat-filled tenders to carry sufficient fuel to generate enough steam for the journey. It was trialled in Scotland, but never got off the ground commercially.

On a volume basis, it takes six times more dried peat to produce the same amount of energy as coal, so in terms of the economics of space and heat there was only one winner. Although peat was cheap, it was bulkier to transport than coal, had to be dried, and its production was seasonal and dependent on the local weather. As Carol Quinn, Head of Archives at Irish Distillers, explained, 'After the early nineteenth century legislation in Ireland encouraged the development of large still sizes, they simply couldn't be heated to the required temperature using peat. Most Irish distilleries from the nineteenth century on were located in towns or urban centres, far away from peat sources, but very convenient for the importation of coal.'

In places where forests were plentiful, wood had also been used in the past to direct fire stills. In a *Report Respecting the Scotch Distilling Duties* published in 1798, a former distiller at Blair Athol said that his stills were direct-fired by wood and peat, never coal. Other Perthshire distillers were using coal – clearly at the time you used what was local, cheap and in plentiful supply. The traditional use of wood still lives on in the new Shizuoka Distillery near Tokyo in Japan[12] and in the wood-fired mobile distilleries that tour the small farms of southern France in the autumn and winter, making Armagnac from grapes and *eau de vie* from the fruit of local orchards.

The Romans were the first to burn coal in Scotland 2,000 years ago, after which it was mined on a small scale at the surface through small holes called bell pits. Up until the end of the 1700s, Edinburgh maltsters and brewers collected 'sea-coals' from the shores of the Firth of Forth. On Boxing Day 1760, the Carron Company started up a coke-fired blast furnace near Falkirk. Huge amounts of coal were required to work iron ore and make a range of products, including a naval cannon called the Carronade, which was so devastatingly effective at the Battle of Trafalgar that Napoleon christened it the 'murder gun'.[13] The company later went on to make the famous red post boxes and telephone kiosks that once covered the British Isles. With the invention of the steam engine, coal went on to fuel Scotland's industrial revolution and by the 1880s, there were more than 800 working mines. When production peaked in 1913, just before the start of World War I, mineworkers and their families made up 10 per cent of Scotland's population.[14] Nationalisation of the UK's 1,000 coal mines on 1st January 1947 created the National Coal Board (NCB) and heralded a period of investment and expansion that led to the development of the so-called 'super pits'.

More than forty years after my first visit as a young geology student to Lady Victoria, I returned to what is now the National Mining Museum of Scotland to meet two ex-miners, James Burns Hogg and Neil Young, to talk to them about their memories of the past. Lady Victoria Colliery at Newtongrange, just outside Edinburgh, was built at the height of the Scottish coal-mining era and operated for 110 years before closure in 1981. Deep coal was the prize and during its working life, the mine brought 40 million tonnes to the surface.[15] At the time, it was a huge modern and sophisticated mine and when we visited in 1976, 1,000 miners were employed by the NCB to work underground and at the surface to process the coal. Over a cup of tea, James and Neil, who now guide visitors round the museum in their orange miner's overalls, talk about the working mine as if their last eight-hour shift had just ended. At the pit face under the protection of the hydraulic supports, they worked in temperatures of 30°C and high humidity. As the line of the self-advancing face moved forward and the continuously driven belt took the coal away at their feet, the roof collapsed behind them. Sometimes it didn't, and as James put it, 'a space as large as a football pitch would form that might take days to come down. When it did, it sucked the air out of your lungs; it was the worst noise a miner could hear.'

At one time, 120 pit ponies worked below ground, before being replaced in 1925 by the endless-rope haulage system that transported the miners to the coalface. Rats were a constant companion of the miners, originally attracted by the food for the ponies: 'When the pithead was finally sealed, the rats were trapped and died in the pit.' In the humidity and darkness of the mine, large fungi grew on the rotting wooden pit timbers and James remembers knocking off the fruiting bodies from the walls of the tunnels as the bogie trains sped through the labyrinth of tunnels.

Coal from seams with names like Diamond, Coronation, Sillie Willie and Kaleblades was washed and sorted above ground, loaded into wagons and dispersed along the railway systems of Scotland and beyond. Railways connected the Lady Victoria mine to markets in Edinburgh and Leith as well as the Borders mill towns, local paper mills and a new breed of coal-fired power stations. Locked away in the archives at the National Mining Museum in Scotland are transit labels that were displayed on the sides of coal trucks showing the destination of the fuel and the name of the buyer. In the 1920s, the Fife Coal Company was sending coal from its Bowhill collieries along the railway system to the Ord Distillery Company (via Dunkeld). Another shows that splint coal from the Newbattle Unit, of which Lady Victoria colliery was part, was destined for Rosebank Distillery near Falkirk. The Central Belt was effectively powering distilleries up and down the length of Scotland.

One of the first questions a distillery tour guide will ask is what are the three ingredients required to make whisky? After we have all answered 'barley, yeast and water', the tour can begin. However, there is a fourth essential component – power, and without the fuel reserves to produce power, Scotland would not have the whisky industry it has today. Its diverse and plentiful fossil fuel resources provided the basis for the industry to expand from its small beginnings, when it was largely dependent on peat, to become the biggest whisky producer in the world and power a boom in distillery size and number that has now reached more than 140.

For two towns, Campbeltown and Brora, a special historical relationship exists between their local fuel source and whisky making. Both are blessed with remote and unusual coal deposits that played a significant part in the original decision to build large distilleries on the remote coasts of Scotland. In time, they went on to make a style of whisky that would become famous throughout the world. Although coal exists on the north coast of Arran and was worked in the eighteenth century to provide fuel for local lime kilns, salt production and the *'sma' stills*,[16] just over the water on the Kintyre peninsula is the much larger Machrihanish coalfield. It was central to the production of whisky in Campbeltown, a town that in its day boasted thirty-four working distilleries; large enough in its own right to be named an official whisky region. The second town, Brora on the Caithness coast, is home to Scotland's most northern coal deposit. A Highland coalfield, a curiosity, it is geologically much younger than the coalfields of the Central Belt and belongs to the Jurassic age, the time of dinosaurs. The carbon-rich rocks of the Brora coalfield formed 165 million years ago, at the same time as the oil and gas fields of the North Sea. It provided the fuel that gave birth to an iconic and much-loved distillery.

But the important coal deposits located in the Lowland Central Belt that fuelled Scotland's industrial revolution were formed 300–360 million years ago in the wet swamp lands of the Carboniferous period. Sir Archibald Geikie called them the 'Great Coal-fields' of Scotland,[17] and from them a transport network spread rapidly, supplying fuel and stimulating the growth and expansion of a power-hungry whisky industry, confident in its future.

Coal Ships and Coal Trains

James Watt's steam engine had powered a revolution in transport as well as manufacturing and with the help of a pier, platform, railway siding or canal wharf, coal could now be delivered cheaply and punctually to the more remote parts of the country. First came the canals, which opened in stages in the late 1700s and early 1800s, and then came a boom in railway construction that in 1863 reached deep into the heart of Speyside and brought coal to its distilleries. The canals, shipping routes and railways of Scotland became the arteries along which coal and raw materials began to flow with increasing frequency to coastal and inland distilleries.

'Railway mania raged' in the latter part of the nineteenth century.[18] Maps of the railway network and the location of distilleries show just how important these new railways became for the whisky industry.[19] Peat was still needed and in the 1830s, coal-fired steamships were regularly carrying cargoes of peat from Stornoway on the Isle of Lewis to the Port of Leith, which were sold at a price of 13s 6d (67p) a ton, destined for the malting kilns of Edinburgh.[20] Irish peat was brought across the Irish Sea to the distilleries of Glasgow and on a November day in 1886, four schooners arrived in Campbeltown Loch – the largest shipments of imported peat received in the distillery town.[21]

Coal-fired steamships were now carrying large quantities of peat across the world. Melbourne with its huge port facilities was a southern hemisphere hub for distilling, and according to Australian whisky historian Chris Middleton, 'There were quantities of peat imported from Scotland by Australia's three largest whisky distilleries to compete head-on with Scotch whisky. Between the 1890s and the early 1980s, I would guestimate this could have been as much as 6,000 tonnes of peat in total.' The Federal Distillery, established by the Joshua Brothers in 1884, received shipments of peat from Islay and the Scottish mainland. It was a symbol of 'Marvellous Melbourne', at the time said to be the richest city in the world. With four large pot stills, two malting floors, a cooperage, grain store and an enormous 40ft² malt kiln, it was one of the largest distilleries in the southern hemisphere, with an annual production of over 4 million litres of spirit in the 1900s.[22] In 1928, DCL built a new distillery at Corio Bay, importing seventy-five tonnes of Scottish peat each year to make a heavy, smoky Scottish-style whisky on the other side of the world. Gilbey's, of gin fame, arrived in the late 1920s and began making whisky after World War II, importing Scottish peat from the late 1940s. At that time, the Gilbey brothers, Walter and Alfred, owned several Speyside distilleries that made a lightly peated Highland-style whisky. Making use of the newly opened Suez Canal, coal-fired steamships had become trans-ocean carriers of peat.

Back home in the Western Isles of Scotland, the Clyde-built puffer with its characteristic red and black funnel became, just like the railways, a symbol of progress in a new industrial age. Until the engine design improved, the puffers were like some noisy aquatic steam locomotive: with each piston stroke, a puff of steam shot skywards from the funnel. Hundreds of these stumpy little coal-fired steamships operated from the 1850s for more than a century, roaming north to south, east to west, up and down Scotland's remote coastal communities, and penetrating inland along its canals and waterways.

Before Kilchoman was built in 2005, all of the Islay distilleries were located on the coast, reflecting the importance of the sea as a route for commerce and survival. All of the smaller inland farm distilleries had failed throughout the course of the nineteenth century due to their inability to compete in a growing industry. The names of the puffers and their association with individual Islay distilleries became the stuff of legend. From 1923, the *Pibroch* brought barley, coal, empty casks and even a piano for the distillery manager's wedding to the piers of Caol Ila and Lagavulin, before being replaced by a diesel-powered ship of the same name. Twice weekly and weather permitting, the *Texa* brought barley and coal to Caol Ila from Broomielaw in Glasgow and returned with whisky for the bonded warehouses, bottling halls and blenders of the city. In the summer months, the *Texa* was requisitioned to carry distillery peat cut from the Bog of the Bulls behind Bunnahabhain Distillery along the three-mile stretch of coastline to the pier at Caol Ila, where it was unloaded and stored in the peat shed.[23] When Alfred Barnard visited Bowmore in the mid-1880s, it had its own steamship, *James Mutter*, named after one of the distillery owners. But the most famous Clyde puffer of them all was captained by Para Handy. The *SS Vital Spark* and its merry crew puffed their way around the coastline and piers of the Western Isles and in *A Drop O' the Real Stuff* even got up to a bit of whisky smuggling in Arran.

Just how much of a lifeline the puffers were is catalogued in the Bunnahabhain Coal Books[24], which list the tonnage of coal received, the names of the puffers and the number of loads that arrived at the pier each year between 1928–63. Each puffer would deposit around 110 tonnes at the distillery before heading back south to the coalfields to return with more. Names like *Dorothy*, *Spartan*, *Anzac*, *Tartar* and *Druid* are recorded time and time again on their frequent trips up and down the Sound of Islay. When in production, the distillery was using 50–60 tonnes per week and in years of plenty, received up to twenty-two puffer-loads of coal a year. Shortages of coal were greatest in the final years of World War II and during the economic recession that followed.

With the arrival of regular coal supplies, distilleries could use new ways of generating power and better control process, either by directly firing the stills or indirectly by the production of steam in large coal-fired boilers. Coal was also used to power the little shunting engines in distillery sidings. At the end of the 1800s, John Duff built a private railway to connect the half-mile between his two distilleries on the Spey and for many years Puggy, his private steam locomotive, puffed back and forth between Benriach and Longmorn carrying barley, coal, peat and whisky. As the railway system expanded, new distilleries like Ardmore and Knockdhu sprung up, and others like Balblair were moved and rebuilt closer to the network. The construction of Ardmore in 1898 is a clear example of a distillery built using the ability of the railway to transport materials to the construction site, and when it became operational, provide it with its fuel. In a testament to the age, it was the largest coal-fired distillery in

Scotland with eight stills that operated until 2002. When in 2005 the beautiful brick coal furnaces below the stills at Glendronach finally went cold, it became the last distillery in Scotland to end its dependency on coal.

The 1970s and 80s were a time of world economic recession, war in the Middle East, a global energy crisis, labour strikes and, as the markets for coal declined, the future for the industry looked increasingly bleak. Coal mining in Scotland, which like the whisky industry had been cyclical in nature with boom following bust, finally came to an end in March 2002. Shortly after the last deep coal was brought to the surface at Longannet Pit in Fife, 17 million gallons of water flooded the productive coal seams 600m below the surface, signalling an end to deep mining in Scotland.

Coal, the Flavour?

So, in the same way that peat smoke found its way into whisky flavour, did the distinctive smell and taste of coal smoke have an influence on whisky? In the malting kilns, coal or coke was used and is sometimes still used to dry barley, in combination with peat or on its own. Coke is a form of clean coal produced by heating coal to a high temperature in the absence of air to remove impurities and increase its carbon content and calorific value. In effect, coke or anthracite was used to create the heat to dry the malt, whilst peat was used to create the smoke and flavour the malt. More than 100 years ago, J. A. Nettleton observed under his microscope drying malt containing 'numerous bright or dull specks of coal dust or coke dust' drawn up into the grain from the 'air-tight kiln' below, with the result that 'the taste will be smoky'.[19]

Depending on the quality of the coal, the smell and taste of its smoke can be variously described as dry, acrid or slightly acidic, dirty or earthy. To some, coal smoke can taste much like the stub of a cigar or smell like the sulphurous end of a half-burnt matchstick.

Much of the coal burnt with or without peat was anthracite ('blind coal'), the cleanest, most pure form of coal. However, it was extremely rare in Scotland and expensive, as it had to be transported north from the deep coal mines of South Wales. Sometimes marketed as 'naturally occurring smokeless fuel', it burns with almost no odour and was attractive for that reason. Coal quality was less of an issue for distilleries with large coal-fired boilers that consumed huge amounts of low- or medium-quality coal to produce steam.

Peatiness in whisky is often described with words like pungent, rich, earthy, damp, in contrast to smokiness that conjures up dryness, burning wood, beach bonfires. From a time when coal was more widely used by the industry, its smoke crops up regularly in the tasting notes of old malt whiskies from certain well-known distilleries. It has been detected in various vintages such as Glen Grant 1949, Springbank 1974, old Bowmore and Knockdhu, and surprisingly in a Longmorn from the 1930s. A Glen Grant 1948 bottled by Gordon & MacPhail has coal smoke in its sensory profile after a period of seventy years in a cask.[25] For some, coal smoke is part of the taste and nose of 1970s Brora... 'becomes tantalisingly gamey and more smoky as it develops, like eating a mutton pie on a coal-fired paddle steamer'.[26] Not an everyday experience, but maybe a common case of phantosmia, or olfactory hallucination, a common condition associated with smelling odours that are not really there.

As the use of coal died out, it is surprising to still find the taste and smell of coal in more recent vintages from the 1990s onwards, appearing regularly in modern tasting notes of several whiskies including Ardmore, Cragganmore, Caol Ila, Lagavulin, and Mackmyra from Sweden. I asked Angus MacRaild if this is simply a form of auto-suggestion, finding smells and tastes that are expected to be found from a name and a date:

Yes, without doubt there is a level of auto-suggestion involved in tasting and nosing whisky, but with Ardmore I regularly get funk and coal on the nose. You can find these characteristics to varying degrees in today's Springbank and some older batches of Caol Ila from the 1970s and early 80s. There is a distinct lack of research or understanding around what, if any, flavour qualities coal lends to whisky via the malt-kilning process. As such all we have is historical anecdote and our own subjective impressions. Perhaps what it really suggests is that peat is a profoundly influential and important ingredient in Scotch whisky when used.

Less surprisingly, coal is part of the taste profile of Yoichi Single Malt, produced by a Japanese distillery that still uses direct coal-fired stills. But there is one whisky region in Scotland that has always had coal firmly at the centre of its taste profile.

Campbeltown Soot

When travelling down the Kintyre peninsula with Islay on your right, the rocky hummocks and metamorphic mountains of the Highlands suddenly open out into the expansive flatlands and green fields of the Laggan. It is a scene more reminiscent of parts of Scotland's Central Belt and I am back in the coal-bearing rocks of the Carboniferous. The coast road has taken me past old raised beaches, grounded sea stacks, fossil caves and cliffs, all evidence that sea levels were at one time much higher. At the time the Laggan was submerged, and as the seas gradually retreated, these wet, poorly drained flatlands became the perfect place for a sandy, salty peat to form. Coal, peat, fertile agricultural land and water – the area was blessed with all the natural attributes to make whisky.

I drive on through the rain into the Laggan, along Moss Road and past tall military fences that enclosed a strategically important

RAF and NATO airbase where nuclear weapons were once stored. At one time, it had the longest runway in Europe, and more recently was earmarked as an emergency landing site for the NASA space shuttle. It was home to the huge delta-winged Vulcan bombers that operated during the Cold War and carried a nuclear deterrent, but now it has morphed into Campbeltown Airport, a quiet airfield with just two flights a day to nearby Glasgow. The runway is eight metres above sea level, but as I drive round the perimeter fence, it somehow seems lower. On the ground, I find no signs of peat banks or harvesting operations; the peat has long been removed and the land reclaimed. In its place are green fields and productive arable land amongst the isolated farmsteads dotted across the Laggan. Rhion Farm was once the site of much peat cutting for Campbeltown's distilleries. I drive past Aros Moss, another major source of Campbeltown distillery peat, now just a wet sunken depression of open water pools, head-high bull rushes, willow scrub and birch. This is a very wet place covered with a geometric pattern of deep intersecting drains full of peaty brown water that hardly seems to move. Most of the moss peat is inaccessible behind the perimeter airfield fence, now out-of-bounds. Much was removed and replaced by tarmac during construction of the runways and taxiways, but the patterns of old peat banks and drainage systems are still clearly visible from satellites looking down from above.

On the road to Machrihanish, I pass through the former mining town of Drumleman. All that remains is a row of cottages facing the Miner's Mission Church across the road. Coal was mined here from the 1600s and maybe even earlier. It was poor-quality coal, mostly sold locally, but at times was shipped to the coastal ports of Ireland, with Irish peat returning on the homeward crossing when supplies at the Campbeltown distilleries were running low.[27] At its peak, the coalfield employed 400 miners, but working conditions were hard and salt-water ingress and flooding was a constant hazard, leading to loss of life on several occasions. High-quality coal, shipped by coastal puffers to the farms and towns of Kintyre from the Ayrshire coal fields, was often used to supplement the inferior local supplies. After periods of dormancy, the sinking of new pits, changes in ownership and nationalisation, local coal production finally ended in 1967. It is estimated that the Machrihanish coalfield contained 70 million tonnes of coal – much of this rich deposit still lies below ground or beneath the sea. Its quality may have been inferior, but the presence of a local coal deposit had a significant influence on the style and taste of whisky made in Campbeltown.

Peat and coal from the Laggan were the energy sources that powered whisky making in Campbeltown, but when a three-mile-long coal canal was constructed in 1794 connecting the coalfield to the town, it triggered a significant shift in the balance of power between the two fuels.[28] Centuries of peat cutting on the Laggan was now threatened. Coal began to out-compete peat, although there was still work for the peat cutters and in one twelve-month period in the 1840s, 4,820 cartloads of peat were driven to Campbeltown, almost entirely cut from the Laggan Moss. Coal became the fuel of choice for the better-off middle and upper classes of Campbeltown, although the poorer folk still relied on peat for heating and cooking. Whisky production in Campbeltown had already passed its peak in 1875, but it was still a large industrial town of 8,000 people with twenty working distilleries, hungrily consuming 600 tonnes of coal a week. The distilleries remained good customers for peat and relied on individuals like John MacKay who in the early to mid-1900s delivered dry peat cut from the Aros Moss to most of the town's distilleries, including Benmore, Glengyle and Lochhead.[28] May and June were the best months to cut peat and John would produce about 200 tonnes or 400 cartloads a season.

PEAT AND WHISKY

* * *

The curtain of rain moves away east and as the afternoon brightens up, I head to Campbeltown, once home to thirty-five legal distilleries – now just three remain. For better or worse, real or imagined, coal has always been at the heart of the taste of its whisky. The smell and taste of Springbank single malts has been variously described as 'coal dust, dry smoke, earthy smoke, rugged smoke, a hint of coal, tobacco, coal smoke and Campbeltown soot'. Alfred Barnard[1] was famously unenthusiastic about its whisky, calling it 'generally thin, useful at the price'. Campbeltown whisky, including Springbank, was used by John Walker of Kilmarnock in his early blends; the heavy, west-coast peaty flavour much favoured by toddy drinkers of the time.[29] However, the 'heavy and robust' style created by the use of peat and coal was not to everyone's liking and by the turn of the century, blenders were beginning to look elsewhere. In response, some distilleries attempted to modify their style of malt whisky: 'The turn of the twentieth century brings a change of whisky preferences and Springbank alter their production accordingly to make lighter whisky that was not as heavily peated, using coal rather than peat to dry malt.'[30]

The pioneer of Japanese whisky Masataka Taketsuru spent the first few months of 1920 at Hazelburn Distillery and wrote[31] that 'a mixture of Welsh smokeless coal and peat' was used in the drying kilns. He then goes on to highlight the importance of Welsh coal, writing, 'it is said to imbue the malt with its characteristic whisky aroma'. At this stage in his research, he had begun to conclude 'that it will be extremely difficult to replicate Scotch whisky in Japan, where we are unable to produce peat or Welsh coal'. He was right about Welsh coal, but wrong about peat.

Coal is no longer used by the distilleries of Campbeltown, but the historical changes in the use of peat and coal created different styles of whisky, which to this day are reflected in the core range produced at Springbank: Longrow, a heavily peated spirit first made in 1973; Hazelburn, an unpeated malt created in 1997; and the 'traditional' light, subtly peated Springbank. The old coal furnace still sits below the large wash still that is now fired by oil with additional steam coils. Peat has outlived coal at Springbank, but it is no longer sourced locally. Supplies became more and more difficult to come by, and two of the last distillations to use local Laggan peat were the famous Springbank Local Barley vintage expressions of 1965 and 1966 where, 'with the exception of the oak cask, all materials were sourced from within an eight mile radius of the distillery'. On the back of the bottles are the words 'barley from the Machrimore Farm malted at the distillery using peat for drying from the Aros Moss at the Rhion ... The coal used for heating the mashing water and the stills was mined from the Argyll Colliery at Trodigal near Machrihanish.'

The maltings at Springbank closed sometime between 1967 and 1977, and when Frank McHardy went to the distillery in 1977 as manager, they had fallen into a state of disrepair. The maltings were re-opened in 1992 and the last time local peat was used to fire the kiln at Springbank was at the very end of the twentieth century. It was coastal peat cut from the Laggan, sandy and impure as well as expensive, so the distillery went looking elsewhere. When heavily peated Longrow started to be made in the 1970s, the distillery was already using machine-harvested peat from Tomintoul in the Highlands.

Alfred Barnard clearly enjoyed his two-week stay in Campbeltown, taking his time to tour its twenty-one distilleries. The harbour was 'teeming with life and hundreds of sail' and when his party stepped out from the White Hart Hotel one hot summer day on their way to Glen Scotia, they encountered 'many hardy fish women with sunburnt faces, selling fresh herrings which glistened like silver in the sunshine'. This morning, it's a little different, and although I didn't encounter women selling fish on my way to Glen Scotia, I did meet someone at the distillery gates who looked like a distinguished bearded mariner. I didn't recognise him at first, but it turned out to be Iain McAlister, the distillery manager.

In the late 1880s, Glen Scotia had three kilns 'heated with peat and mixed with a little blind coal', but with the end of malting in the 1950s–60s came the end of peat use at the distillery – all their peated malt is now sourced from Glen Esk maltings on the east coast of Scotland. I asked Iain if he could describe to me what makes Campbeltown whisky different: 'Complex, robust, oily, with a touch of salt', he answers and then pauses while drawing an imaginary line in the air, 'and a thread of peat running right through it.' There is more: 'and a touch dirty. Campbeltown in its heyday was a place of coal fires, with soot and smoke in the air – you can still see the blackness on the walls of the older houses and tenements. Campbeltown was, and still is, a working town.' That is all too evident when you walk through the streets; old and new, restored and derelict, dark and light, weathered and pristine. When I visited at the end of 2021, the town was alive with rumours of three new distilleries in the air. The rebirth of Campbeltown whisky was gathering pace.

The link between Campbeltown and Japanese whisky is well known, but almost fifty years before Masataka Taketsuru visited Hazelburn, a distillery manager from Sweden[32] called Gottfried Olsén came to Glen Scotia Distillery for a week in 1872. He worked for the Göteborg Jästefabriks AB in Sweden, the Gothenburg Yeast and Spirit Factory, and after spending time at Pulteney and Millburn distilleries, he travelled south and west to the 'Victorian whisky capital of the world'. Like Taketsuru after him, he was on a fact-finding mission to learn how to make peated, maritime whisky. When he returned to Sweden, he used that new knowledge to produce Sweden's first whisky in the 1880s. No bottle has ever been found, but we can safely assume that it was in a style that would be familiar to those of us that enjoy lightly peated, salty maritime malts.

Iain introduces me to local historian and writer Angus Martin, a person who knows everyone and everything about every building in the place. Tweed bonnet, tweed jacket, tough boots and a weathered backpack, he has the air of someone who has tramped all the paths of Kintyre unearthing its history and recording its stories, and then tramped them all some more. He digs into his old backpack and gifts me one of his books, *Kintyre Country Life*. In it, I later read stories of peat cutters, whisky smugglers and their encounters with excisemen. Illicit whisky was a profitable and an important source of income for many in Kintyre, and when Thomas Pennant visited Campbeltown in 1772, he wrote that the 'locals being mad enough to convert their bread into poison, distilling annually 6,000 bolls of grain into whisky'.[33] Peat fuelled the stills of the whisky smugglers who operated in secret town hideaways, remote farms, glens and sea caves. Some of the whisky makers were natives of nearby County Antrim, operating in Islay as well as Kintyre, and in the early 1800s, illicit distillers could buy barley directly from malthouses in Campbeltown. Excisemen carried guns and lost their lives in their line of duty. 'The stills are generally discovered by the smoke, though expert Excisemen also trace them by examining the water of the mountain streams impregnated by the waste.'[34] The last illicit whisky made in Kintyre was around 1880. Angus walks everywhere, has no car and buys his whisky within a tight budget. He is working on a new book and hopes to see it published, 'if I survive that long'.

The story of Campbeltown whisky and its signature style is one that reflects not only the place, but two rival and competing sources of fuel. During its golden age, coal was always going to be the winner over the 'miserable and precarious kind of fuel' that was peat. With the benefit of time, that victory can be seen in the context of what followed.

Fields of Oil and Gas

The golden age of peat may have been ended by coal and steam power, but the latter part of the 1800s saw the arrival of a new fuel that changed the industrial landscape forever and triggered violent conflicts that continue to fester in the world to this day. In 1851, the world's first oil refinery was opened by James Young at Bathgate to the west of Edinburgh to distil paraffin oil from a brown-black rock with an oily smell called cannel coal.[35] His earliest oil stills were made of thick cast iron, fired directly by coal or coke that evolved with time to become fat, nine-foot-high pot stills. The Scots called it 'candle coal' as it burnt with a large bright yellow flame. Most people now know this as oil shale, a fragile rock with a rubbery or greasy feel that occurs stratigraphically below and is older than the Carboniferous coal measures. James Young's method of distillation kick-started more than a century of oil shale production in Scotland that ended in 1962 – all that remains now are the rusting red, and sometimes toxic, heaps of oil shale waste that form the landmark bings for travellers heading west out of Edinburgh. One former oil shale mine site at Bathgate is now occupied by a massive grain distillery called Starlaw, a French-owned operation. It seems very likely that oil produced from Scotland's oil shale fields was once used to either fire stills directly or generate steam for Scotland's distilleries. The petroleum it produced was once sold at characteristic thistle-shaped pump heads and marketed as 'Scotch'.[35]

Eight years after James Young opened his refinery, Edwin Drake struck oil on the other side of the Atlantic Ocean, 21m below ground at Titusville, Pennsylvania, and became the first person to successfully drill for oil. At the end of the nineteenth century, the world was entering a new energy age, although it was not until well into the twentieth century that oil began to be used by the whisky industry. In the 1930s, direct firing of malt kilns with peat, wood or coal was still commonplace, and just after the end of World War II, many pot stills were direct fired by coal, as they had been for more than 100 years. Direct firing of stills by oil wasn't a huge success and as Bill Craig, former general manager at Allied Distillers, recalled:

In the late 40s and early 50s, some distillers started burning oil beneath their stills as a substitute for coal. It wasn't a success – the high sulphur content of the oil played havoc with the copper.[36]

In 1922, Cardhu on Speyside had become the first malt whisky distillery in Scotland to experiment with heating their stills with oil-fired steam. The experiment only lasted a year – oil was too expensive and the distillery quickly reverted back to coal.[37] The stills at Craigellachie were also for a time direct-fired with sulphur-rich heavy fuel oil, a taste that is still very much part of the character of its whisky.[38]

By the 1960s, oil prices had become low enough for it to be more widely used to generate steam. But conflict in the Middle East in the early part of the 1970s caused a surge in global oil prices that triggered a quest for alternative forms of energy. Natural gas began arriving onshore from the recently discovered hydrocarbon fields beneath the North Sea, and was quickly adopted as a cheap, clean sulphur-free fuel for the direct firing of stills. In the 1970s, electricity became more widely used by the industry, either generated from coal, oil or gas. So, after a long period of stability when coal had been the fuel of choice, the end of World War II marked the start of a period of huge change in fuel use by the whisky industry. Significantly, energy production and usage were becoming cleaner, greener and more efficient.

The Electric City

It was the first Highland town north of Inverness to have a public electricity supply, and in celebration the locals fittingly renamed their home the 'Electric City'. It now has a successful football team that plays in the Highland League, a pair of iconic distilleries and an industrial heritage built on the most northerly deep coal mine in Britain. Almost all traces of the mine have long gone, but you can still find small pieces of soft, dull black coal along riverbanks or at low tide on the beach. Rounded into pebbles over time by currents and waves, light in weight, porous and tinged with brown, they are more peat-like than coal-like. At the Heritage Centre, a modern structure built above the old mine, visitors and local schoolchildren can eagerly explore the town's mining history and hear stories from the past.

Brora coal was 'fiery' and when burnt gave off a distinctive 'vegetable odour'.[39] It was famously prone to spontaneous combustion, either below ground, above ground in waste heaps, or on one occasion on a cargo ship bound for Portsoy on the Moray Firth. It was so sulphurous, stony, combustible, badly faulted and hazardous to work that the NCB avoided mining it, leaving the task to private enterprises. But it was plentiful, occurred in seams with an average thickness of one metre, and was burnt locally for almost 400 years to provide energy for homes, salt production, a large and nationally important brickworks, a woollen mill, a brewery and a distillery at Clynelish. Brora coal was shipped further afield to supply homes and power stations in Inverness, Aberdeen, Thurso and Wick and in the mid-1950s, a 100-tonne load of coal was shipped to the Faroe Islands. In 1913, steam produced from boilers fired by Brora coal was linked to an electricity generator and finally the streetlights of the Electric City burst into life. In the 1950s, the owners of Clynelish Distillery were less keen to use the local coal as it was of poor quality and 'would not raise steam quickly enough', but the quality was improved by the production of cleaner coal briquettes. The Brora Mine, which in its final years spawned a second pit, was worked periodically from 1529 to 1974 whenever there was a favourable economic tailwind. It closed due to a combination of rising costs, shortage of labour, unfavourable economics and safety concerns as the mine became deeper and more hazardous. Records from a mine logbook show that in 1972 it was still supplying large amounts of coal to the Kincardine Power Station on the Forth, with an additional 730 tonnes sent to unnamed distilleries.

Twenty-five kilometres offshore from Electric City is a clue to why Brora coal is so different. Visible on a clear day out in the Moray Firth are the three remaining platforms of the Beatrice oilfield, which since 1980 produced thousands of barrels of oil a day from Jurassic rocks two kilometres below the sea floor. Decommissioned in 2017, the field has become the home to almost 100 wind turbines – the Beatrice Offshore Windfarm, providing power for almost half a million homes. Brora coal is the same geological age as the oil from Beatrice, and sandwiched between the layers of coal are occasional bands of oil-rich shale. Compared to the much older Carboniferous coal deposits in the south of Scotland, the young Jurassic Brora coal was more variable, less compact, less pure, and much more challenging to mine.

In many ways, the story of Brora, its colliery and two distilleries encapsulates the very essence of the story of peat and whisky. The first distillery at Clynelish was founded in 1819 and grew out of an illicit past. Even in its early days as a two pot still farm distillery, it was producing a much sought-after, premium-quality peated single malt whisky. Its malt kilns were fired by peat and its stills by coal. It was purposely built close to the local coal mine, but at times the quality of the local coal produced was so low that it was not used by the distillery. To get around this problem, local coal was mixed with better quality coal brought in from coalfields further south. Peat was cut from a moor close to the distillery, the same moss that provided the source of water to drive the distillery waterwheel. It was significantly enlarged in 1896 and electricity was installed in the early 1960s when the two stills switched from direct coal-fired to internally steam heated. The distillery's dependence on coal ended in 1969, after which the old coal-fired steam boiler was converted to oil.[40]

In February 1968, floor malting ended at Clynelish, the empty peat shed was recycled and the old Clynelish Distillery closed soon after a modern large distillery was built next door for DCL's blends with the same name. But the story of the old distillery didn't end there. After a drought in Islay in the summer of 1968 and a significant shortfall in the amount of heavily peated whisky being produced, DCL made the decision to reopen the mothballed distillery and call it Brora. When it resumed production in 1970, the Brora Distillery aimed to produce an Islay- or Talisker-style single malt whisky, using malted barley from Ord Maltings peated to 45ppm phenols. Peating levels were at their highest between 1970–3. In one special year, more than twice the normal amount of peat was used in the Ord malt kilns. The maltsters and whisky makers didn't know it at the time, but they had created what was to become the most famous Brora single malt vintage of them all. Brora 1972 was probably the peatiest of all the whiskies produced outside Islay and certainly on the mainland. By 1973, things on Islay had returned to normal and with a new, larger and renovated Caol Ila Distillery producing large amounts of peated malt whisky, peating levels gradually decreased at Brora to around seven ppm

– the norm for a Highland distillery. Along with a host of other DCL distilleries in the Highlands, Brora seemingly closed for the final time in March 1983 deemed 'surplus to requirements'.

For all the romance surrounding the old Clynelish Distillery, it does have a dark side. Karl Marx likened it to 'robbery of the common lands',[41] and to most of us the Highland Clearances conjure up images of cruelty, hardship, forced eviction, fragmentation of a language, culture and way of life. Two people will forever be linked with the Clearances – the Duke and Duchess of Sutherland – and it was their estate that invested heavily in a new farm distillery at Clynelish, close to the Brora coalfield. They planned to use the new workforce recently arrived at the coast from the inland townships as distillery workers, miners and farmers to provide sufficient barley for the malting kilns. For this reason, Clynelish, like Talisker on Skye, has been called a 'clearance distillery'. Brora was a planned village and the cleared tenants were allocated lots along the coastal plain, which are clearly visible today at Achrimsdale, East Brora, East Clyne and Dalchalm.

The opportunity to taste a timeline of Brora single malt whiskies through its famous peated period and beyond rarely comes along. On a sunny day in May 2019, 200 years after the birth of the distillery, I joined a celebratory tasting of five Brora single malts from the Diageo Special Releases in the cask-filling store of the old distillery. As a memento, we were all given a symbolic bottle of ash by Ewan Gunn, a local Wick man (or Wicker), but more widely known as a Diageo Global Scotch Whisky Master. It is not often you are presented with a bottle of ash and I for one was overjoyed. Not any old ash, but a sample 'retrieved from the Fire Box of the Charles Doig Kiln at Brora Distillery', fifty-three years after it was fired for the last time in 1965. Now safely in the hands of a curious scientist, I later took a look, a sniff and a taste:

Coarse, gritty texture, an orange-brown sand with a subtle medicinal smell of bandages and antiseptic. On closer inspection – highly variable with lots of creamy white flakes of soft ashed peat, occasional fibrous flakes of black unburnt peat, no coal, lots of round clumps of fused furnace slag, and bright orange grains of fine-grained sandstone. The Fire Box ash suggested to me that the combustion temperature fluctuated with a cool later stage indicated by the unburnt peat. The abundant sand grains and colour indicate the peat used in the Fire Box was impure and cut from the base of a bank above sandstone bedrock. It tasted of the earth.

When Ewan started the tasting in the store with its old 16,000-litre plus wooden spirit vat and peeling white walls, there was an atmosphere of hushed reverence and expectation. The first three drams, which included the much sought-after first Brora Special Release 1972 vintage, were all thirty years old, bottled between 2002–4 and from Brora's heavily peated period. Each one was golden in colour and aromatic, sometimes with dry smoke, sometimes quite pungent with sweetness and fruit, and all having that signature damp, earthy, farmyard Brora nose and taste, with or without the candle waxiness that is characteristic of the modern Clynelish distillate. Once we got into the tasting and started jumping back and forth between samples, the subtle differences between the bottles became clearer. All were well balanced with no one taste overpowering the others. I particularly enjoyed nosing and tasting the 2004 Third Special Release; my style of whisky and in a blind tasting I would have placed this as a true Islay-style whisky, maybe a 20-plus-year-old Caol Ila or Lagavulin. The final two drams were a 35-year-old Brora distilled in 1978 (Thirteenth Special Release) and a 34-year-old distilled in 1982, the sixteenth and last of the Brora Special Releases. We were now into the later vintages that were all less peated and I was stunned by the 1978 vintage – something really special here with peat smoke sitting in the background, subtle but always present, waxy, fruity and spicy notes with liquorice root. Wonderful whisky, a clear winner for me amongst a remarkable array of liquid history. The tasting now over, we looked at each other, stepped out of the old filling station into bright May sunshine, and blinked.

So, what happened to all of Alfred Barnard's 'huge peat sheds' containing '100s of tonnes of odoriferous peat'? He was clearly impressed by peat sheds and of Bladnoch Distillery: he wrote, 'the peat shed is quite a handsome erection'. He found no such shed at Clynelish distillery, only a 'yard for peats'. A huge peat shed did finally arrive only to be deemed 'surplus to requirements' when the floor maltings closed and local peat was no longer required.

Brora Rangers 6 – Huntly FC 0

The kick-off at Dudgeon Park was delayed forty minutes because of heavy rain and a waterlogged pitch. I asked a club steward in his smart red Brora Rangers tie where they got the peat shed from. 'Dunno mate,' he answered blankly. I paid for my ticket, walked through the red turnstile, picked up a damp team sheet, and in the spirit of the occasion bought five ducks for next week's charity duck race on the Brora River. I got chatting to a local farmer and asked him why it was called the Peat Shed. 'Was it named after a man called Pete?' he hypothesised and smiled. As the sound system blasted out *Bitter Sweet Symphony* and *Blue Monday* time and time again and we waited patiently for the match to start, I entered the shelter of the old peat shed, now the East Stand.

I paced it out, five metres wide, sixty metres long, with a new red roof and see-through perspex ends. 'No Alcohol Allowed' read a sign hammered to its side – times have changed both in football grounds and peat sheds. Kevin Innes, the former Clynelish

Distillery manager, remembers 'the lads taking it apart screw by screw and reassembling it down at the ground. That was how things were done then, as a favour to the local community.'

The Huntly fans were out in force, all nine of them. They had their own hypothesis as to why the kick-off was delayed. Apparently, the match day referee was so hungover from the night before that a replacement had to be found. Backs to the North Sea, hands in pockets, we waited for the match to start. As it should be in a peat shed, it was perfectly dry in the squally rain, designed to let the air in and keep the rain out. It was cold in the peat shed and at half-time, with the Cattachs already 3–0 up and accompanied by another blast of *Bitter Sweet Symphony*, I headed towards a closed window to buy hot tea and a sausage roll. As the cold began to bite midway through the second half, I returned again to bang on the now closed steamed-up window to buy a cup of black, comforting Bovril.

For the record, Brora Rangers won 6–0, I didn't win the duck race and Brora Distillery, after being taken apart brick by brick, started making whisky again in May 2021, thirty-eight years after its stills last ran. The 'clearance distillery' is now no longer surplus to requirements.

Four

Unlocking the Smoky Gene

*'The day will come when the blender and the chemist will work
hand in hand, and that art and science combined will achieve
results beyond the reach of either.'*
Philip Schidrowitz, industrial chemist, 1907[1]

Firing of the Peat Kiln

After tea, porridge and kippers, I set off towards the distillery
along the harbour wall in mid-morning November sun. It's low
tide. The smell of rotting seaweed and decomposing sediment fills
the air – solitary gulls waddle across the exposed mud, curlew and
oystercatcher probe for bivalves and further out in the open water,
rafts of eider ducks preen themselves and call to one another. At
the ferry terminal, I read and then step over an inscription in the
walkway by Andrew and James, students at Castlehill Primary
School. They had written 'Welcome to Campbeltown, there is lots
to do in this marvellous town, you could go and see our distilleries,
they sell the finest whisky'.

When I arrive at Springbank, the peat fire is already set. It has
structure – a bottom layer of scrunched up pages of the *Daily
Record*, a thin middle layer of split timber and a top stratum of
dry peat nuggets. On the floor next to the kiln, waiting to be shovelled into the fire, is more dry brown peat and a black wet peat.
In the courtyard outside, a large corrugate-iron shed overflows
with dry St Fergus peat from Aberdeenshire, and close by a separate mound of wet amorphous peat sourced from Bogbain Farm
near Inverness. After the recent rain, it looks like it has just been
exhumed from the ground, more organic mud than distillery peat.

PEAT AND WHISKY

We are standing at the base of the kiln and at 10.52 I am given the honour of lighting the fire. This is the moment when the malted barley in the kiln floor above starts to be infused with the flavours and aromas of peat smoke. It feels almost ceremonial – a key part of whisky making that has gone on at Springbank for nearly 200 years. Three clicks of the lighter and the fire is away. The airflow is turned on to create an updraft and the *Daily Record* is ablaze in seconds, followed by the layer of wood and now the dried peat. The fire doors are closed to get a fast start. John and Roddy are making peated malt for Kilkerran and the requisite six hours of peat kilning has started. The supplementary airflow is switched off and John feeds the fire with more dry peat. After twenty minutes, when he is satisfied the kiln fire has a solid heat at its core, he begins to add the wet peat, skilfully shovelling it along the fire front dampening any flames: 'Smoke is what we want, flame is your enemy.' The smoke rises up into the 'clogie', the smoke-filled chamber beneath the kiln floor, where a one-foot-deep bed of malted barley is starting to adsorb the peat reek. The configuration of airflow in the kiln can be altered to recycle smoke through the malt bed to enhance levels of phenols. Tending the kiln is a real skill and one that is central to the task of making peated whisky at Springbank.

By now, the kiln fire doors have been flung wide open to maintain a natural airflow through the kiln to the brick chimney above. John closely monitors the smoking peat, dampening flames whenever they erupt with fresh shovels of wet peat. 'This is really good peat and it burns superbly, so much so that we are really getting through it quickly,' he tells me. It is not hot on the floor of the kiln room, just warm, pleasantly warm. Our conversation quickly moves on to the kilning of Longrow malt, the most peated whisky made at the distillery. The kilning cycle is entirely different – forty hours of recycled peat smoke behind closed kiln doors, as loaded spade after spade is shovelled to the back, filling the deep kiln. This is intense, physical work, and when John started at the distillery Longrow was made in the hot summer months. The production schedule has now changed to the cold month of January to make the heat more manageable for the operators. John obviously has a passion for making and drinking Longrow – it is his favourite of the whiskies made at Springbank. He points out an area of repointed bricks around the kiln that 'cracked when we were making Longrow. The heat can be so intense that when we clean the kiln after it has cooled, we find beads of glass formed by the melting of sand grains in the peat.' Roddy nods in agreement. When I got home that evening I looked up the melting point of silica: 1,700°C.

On the wall of the kiln room is an instrument that I have not set eyes on for thirty-five years – an ancient metal chart recorder. Two ink pens that record the changes in air temperature as it flows in and out of the kiln are beginning to draw green lines on a slowly rotating paper disc. John shows me the circular chart from the last kiln run – the moment the fire is lit, the rapid rise in temperature to 65°C, the point after six hours when the peat fire is allowed to die and the oil is switched on to generate the hot air to dry the malt at 130°C. He shows me the 'break point' after twenty-one hours when the temperature of the kiln air becomes too high and one of the operators then dons breathing apparatus, climbs into the kiln floor and aerates the malt by digging and re-spreading it. This evens out the temperature in the malt bed.

Roddy arrives with a fresh barrowload of wet peat, smiles and takes over the kiln work as John heads off to the lower malt floor to turn the germinating barley for the next batch of Kilkerran. Eyes smarting and with the smell of Scottish peat reek on my clothes that lingers into the evening, I follow John across the malt floor into the outside world. Peat smoke wafts out of the square brick chimney into the bright Campbeltown air and drops down onto the car park on a rare windless morning on the west coast of Scotland. The neighbours know not to put their washing out to dry on a Wednesday. I leave John and Roddy to their work.

Smoke, the Flavour

Almost everything we eat can be sealed into jars or cans, chilled, dried and frozen, bombarded with highly charged particles and preserved with an arsenal of chemicals, but for thousands of years humans relied almost entirely on salt and smoke to preserve their food. A practical knowledge of how smoke could be used to cure and preserve meat would have spread quickly after early humans started cooking with fire. We need to go back as far as the Stone Age to find the oldest known smokehouse in the world, unearthed by archaeologists in the 1970s near the medieval city of Kraków in Poland.[2] In Scotland, local coal and peat was used to evaporate salty brine in large flat basins or 'pans', and small settlements sprung up along the coast with names like Kennetpans, Prestonpans and Saltcoats. In the Age of Discovery, smoked and salted foods were the mainstay of the Portuguese and Spanish navigators who set sail from the shores of Europe on long and uncertain sea voyages to travel east to the Indies and west to the Americas.

Five hundred years later, there is much less smoke in our lives and we no longer need to use smoke to preserve food. Smoke, the preservative, has now been replaced by smoke, the flavour. The list is long and never-ending: cheeses like *Rauchkäse* and mozzarella affumicata, sausages and hams, smoked paprika, chipotle, olive oil, smoked vegetables, and of course kippers and the untouchable Arbroath Smokie. On the steep slopes of the Douro Valley in northern Portugal, I once watched a beekeeper using wood smoke to stun his bees while he plundered their valuable food store. His thick, viscous honey was powerful, very smoky and not for the faint-hearted. Tasting it months later instantly brought back evocative memories of place – it had real terroir.

Many of us have become specialists in smoke and like to know what has been burnt or if the smoke is real or artificial. This isn't just the domain of middle-class foodies; as far back as 1880, a general store in Norwich was not just advertising smoked bacon but peat-smoked bacon in its shop window,[3] the pork cured by turf cut from the salty fenlands of eastern England. Companies began manufacturing products such as Wright's Liquid Smoke, a pyroligneous acid that could be pasted on to meat and vegetables to make them taste as if they had been smoked over wood.[4] It was probably inevitable that at the end of the twentieth century, research began to appear about the potential carcinogenic effects of wood smoke,[5] which led to a decrease in the production of heavily smoked foods in favour of lighter smoking and the use of artificial liquid smoke. But the appetite for the flavour of smoke persists and continues to grow. There is even a perfume on the market called Metamorphic – created as a homage to smoky Ardbeg whisky – with the enticing aromas of peat, spice and black pepper.[6] And, another homage to Islay's smoky whiskies – an Octomore barbecue sauce created by a well-known Scottish restaurant. You might even want to purchase your own personal Ardbeg barbecue smoker complete with resinous wood chips and safety instructions.

Although the Romans are known to have preserved their wine with smoke,[7] in the drinks industry the relationship with smoke is rather different as it has always been used as an accidental or deliberate flavouring and not as a preservative. In the UK, we can buy bog beer, a dark and rich porter made from peat-smoked malted barley, or beers such as Smog Rocket, inspired by London's dense smoke-fogs (smogs) of the 1950s, caused by the burning of coal in domestic fireplaces, pubs and local power stations.[8] Whisky barrels that once contained peated whisky become infused with a memory of smoke and are used by some beer producers for finishing and to add something different. While many of these innovations might be regarded as gimmicks or the outcome of a desire to create something crafty and unusual, that cannot be said of a German smoked beer. Originating from the town of Bamburg in Upper Franconia, *Rauchbier* has been made since the 1500s from green malt that is dried over a smoky beechwood fire and matured in 700-year-old cellars.[9] The original *Rauchbier*, made at the legendary *Accht Schlenkerla* brewpub, is rich, black and coffee-coloured, not unlike a porter. To my nose it is sweet, aromatic and smoky, with a distinct hint of dampness or green wood, smoky barbecue vegetables and resin. It is bottled at 5.1% abv and tastes surprisingly good with a nice balance between bitter, sour and sweet. We might want to call this a craft beer, but a product that has been successfully made and sold for over 500 years is clearly in a class of its own. Others have followed with varying amounts of success – Loch Lomond Breweries once produced a peat-smoked ale and the Isle of Skye Brewing Company produce the wonderful *Tarasgeir* – the Gaelic name for a peat-cutting tool – using peated malt from the Ardmore Distillery. In 2020, Ardbeg launched The Shortie, a smoky dark porter named after the distillery's famous canine mascot and made from Ardbeg's peated malt barley. The list continues to grow and shows no sign of ending.

At higher alcoholic strengths, the trend for experimentation and new smoky products accelerates. In Ireland you might, or might not, be in for a treat if you buy peat wine made by infusing German red or white wine with peat harvested from West Limerick. In the world of spirits, Burning Barn Spiced Rum, flavoured using smoked applewood, was apparently inspired by a fire that engulfed the family barn in Warwickshire. White spirits like vodka have been made with peated barley, and both vodka and gin have been marketed after being aged in ex-Islay casks that once contained peaty whisky. In Italy, they make a spirit from grape-marc or

Pomace, prepared by smoking the pressed and pulpy remains of the seeds, pips and skins.[10]

Smoky malt whisky and bourbons are now appearing all over the world using a bewildering array of combusted organic fibres to create new flavours: Swedish juniper wood, Danish heather, New Zealand manuka wood and Texan oak, mesquite and apple wood, to name just a few. Iceland's Eimverk Distillery just outside Reykjavik produces *Flóki*, a young single malt using barley smoked from the organic-rich waste of sheep's dung. Tasting it was an interesting experience. Cocktail creators are not immune to the smoke addiction. Smoky Cokey is a fusion of Lagavulin 8- or 16-year-old single malt and cola, which clearly divided opinion at a recent whisky show in London. I once had the 'pleasure' of having an icy mixture of Laphroaig 10-year-old malt whisky and ginger ale thrust into my hand at a pub in Edinburgh by a couple of enthusiastic young brand marketers. I did not let it spoil my evening.

Apart from whisky, the other spirit that has smoke at the heart of its flavour is mezcal, produced in a country where smoke is still seen as a purifier of the soul – something magical, medicinal and spiritual. In Mexico, the core of the agave plant, the *piña*, is slow roasted and smoked in a sunken and sealed earth oven above hot wood embers, charcoal and hot rocks.[11] After anything up to thirty days, the slow-smoked *piña* is retrieved, distilled (often in wooden stills), and drunk as an unaged, undiluted, clear spirit traditionally stored in clay bottles or flasks. The variety of tastes is extraordinary, reflecting both the different species of agave and the different geographical regions of Mexico. Sometimes the spirit is light and gin-like, always floral with ashy smoke, boot polish, leather and green resinous pine on the nose. Mezcal Derrumbes, a small-batch, traditional artisanal spirit, has to me the amazing nose of a pair of freshly nikwaxed boots that have just walked through a smouldering fire of wood ash. Sweet, soft and smoky – it is one of the most original spirits I have tasted. Mezcal is a spirit that is made by village artisans harvesting the agave plant from local hill slopes using centuries-old methods. Everything, including the flasks in which it is stored, 'comes from the earth' – the liquid product, a definition of terroir.

'The most Scottish of whiskies.'

Neil Gunn, passionate Scottish Nationalist, prolific author and one-time customs and excise officer at Glen Mhor Distillery, insisted peat smoke 'constitutes one of the flavours that a perfect whisky should possess'.[12] He is not alone in this view – probably the greatest whisky writer of them all, Michael Jackson, a Yorkshireman and son of Huddersfield, believed that peat smoke flavours and aromas were the very essence of Scotch and what distinguished it from all other types of whisky made around the world. Writing about peated whisky made on Islay, he called it 'the most Scottish of whiskies', and 'theirs is the character that makes a blended Scotch unmistakably Scottish'.[13]

There are different ways of introducing peat-smoke flavours into Scotch whisky, but the traditional and established way is at the first stage of the production process when peat is burnt in a smouldering kiln during the drying of malt barley. In the past, peat was used in the kiln both as the primary source of heat and smoke, drying and withering the barley to the required moisture content prior to milling. This produced the all-important peat reek flavours that have become synonymous with many types of Scotch whisky. In modern distilleries and maltings dry hot air, anthracite or coke now provide the heat, but not the smoke. From the time when whisky makers had just one source of fuel to dry their malt, peat smoke has become the very heart and soul of Scotch whisky.

There are other ways of infusing whisky with smoke, some of them more successful than others. It is easy to imagine how, in the pre-1823 days of the private and illicit still, the pervasive aromas of a burning peat fire would have found their way into everything inside the smuggler's bothy, including the new-make spirit. But it is less easy to imagine the aromas of peat or coal smoke finding their way into whisky made legally in large direct-fired stills where all parts of the distillation process, with the exception of the spirit safe, are completely enclosed. Some Scottish distilleries in the early part of the twentieth century were experimenting with different ways of enhancing the peatiness of their spirit after malting.[14] Distillers tried to increase the peatiness of the sweet sugary mash by placing bags full of broken pieces of peat in the hot liquor water prior to mashing – rather like using a giant peat-filled teabag. They also began placing baskets full of smoking peat embers into the washbacks prior to fermentation. After letting the fermenting vessel fill with a dense cloud of peat reek, the basket was removed and the washback filled with wort. Although none of these innovations were widely adopted, they were likely inspired by the time when bags of barley were steeped in peaty bog and stream water by the illicit distillers to initiate germination. Even to this day, the sentiment still persists amongst some that naturally peat-rich distillery source water contributes to the smoky flavour of the final whisky.

In North America, charcoal, a pure form of carbon produced by burning peat or more commonly wood in an oxygen-starved atmosphere, is a key part of whisky production, where it is used as a purifier and a way of cleansing the spirit. On the side of a bottle of Heaven Hill Old Style bourbon, the label states 'charcoal filtered just prior to bottling'. Tradition tells us that in 1825, Alfred Eaton invented a unique charcoal leaching process that became integral to the making of Jack Daniel's Tennessee whiskey, although the process may well have been inherited from slave distilling traditions, when charcoal was used to purify illicit alcohol.[13] Immediately after the spirit is distilled, charcoal is used to remove undesirable congeners and fusel oils. This is a long, slow clean that mellows the whiskey and takes place over a period of seven to ten days, during which time the spirit is trickle filtered through a four-metre-thick bed of charcoal made from sugar maple. Some Jack Daniel's drinkers claim they can detect a faint smokiness in their whiskey.[15] This unique charcoal filtration method led to the US Government granting Tennessee whiskey its own generic status. The use of charcoal was not confined to American whiskey, and in Ireland and Scotland at the beginning of the twentieth century it was commonly used as an additive to the stills, again for purification purposes.[16] In the production of white spirits like vodka, charcoal filtration is widely used to purify and remove unwanted compounds.

A modern innovation in flavour is the introduction of peat smoke into whisky during the maturation stage. Spirit made using unpeated malt can be filled into casks that once held heavily peated whisky produced from distilleries like Laphroaig or Ardbeg. This has become popular among some distilleries, like Pulteney and Wolfburn, that want to do something a little different, but only use peat-free malt to make their new-make spirit. This might be for an extended period of time or a short finish, a final burst of woody peat flavour at the end of the aging process just before the whisky is bottled. Cask charring is also a way of getting lighter smoky char flavours into whisky. Toasting the inside of casks at high temperatures synthesises and volatilises phenolic compounds, enhancing the smoky flavour. In Ireland, West Cork Distillery produce a whiskey using peat-charred casks, in effect peating the inside of the barrel instead of peating the malt. They burn local peat inside the barrel using a traditional bellowing technique, before the whiskey rests in the casks for a final six months prior to bottling.[17] And if all else fails, you can resort to the Swiss Castle method – injecting casks full of maturing spirit with smoke.[18] All clever innovations, but there is only one real way of making 'the most Scottish of whiskies'.

Soil Fires and Peat Kilns

In August 2017, news emerged of an unusually large tundra wildfire in an ice-free coastal region of west Greenland near the town of Sisimuit.[19] Over the coming weeks, satellites began to capture dramatic images of the fire growing in size to eventually cover an area of nine square miles as it rapidly became one of the largest wildfires ever recorded. The locals call them 'soil fires' and say that their frequency has increased as the permafrost has melted and the peat, now free of its icy grasp, has dried and become more combustible. In the moist, low oxygen atmosphere below the peat surface, pyrolysis – high temperature combustion in an oxygen-starved environment – takes place, and when the smouldering fire front reaches the surface it bursts into life and flares. They are part of an annual global fire season that covers North America, Siberia, SE Asia and the Arctic regions, when in the dry summer months smouldering peat and coal fires break out below ground.[20] They can grow to become megafires, as the fire front propagates laterally and vertically, making them extremely difficult to control and extinguish.

As far back as the nineteenth century, explorers reported seeing peat fires in the Okavango Delta as they travelled deep into the interior of the African continent.[21] In tropical SE Asia and particularly Indonesia, peat fires have been reported since the 1960s, many of them started on purpose as a part of a form of cultivation known as 'slash and burn'. Some fires have been known to smoulder for years before they eventually reach the surface and in 1997, the El Nino climate event triggered widespread peat fires across the island of Borneo.[21] The world's longest living subsurface fire still smoulders away at the well-named Burning Mountain in New South Wales, Australia, where a deep coal seam has been smoking for 6,000 years.[21]

Peatlands burn because they are the most fuel-rich ecosystem on the planet. They form part of a thin veneer of flammable organic matter that is wrapped around the surface of the Earth and sits within our oxygen-rich atmosphere. Fuel and oxygen form two sides of what is known as the fire triangle and when that is completed by the addition of heat, the fuel-oxygen mixture ignites. The source of ignition could be a lightning strike, a match, cigarette end or a spark from machinery. No one knows exactly why an explosion occurred at the Dalmore Distillery on 5th February 1919, which completely gutted the peat store, a warehouse and offices, but the chief suspect was a paraffin-driven fire engine owned by the US Navy that at the time was garrisoned at the distillery.[22,23] Fires in dry peat stores were always a constant hazard, but nothing has compared to a fire that took hold and destroyed the entire peat storage capacity of the city of Uppsala in Sweden. The city had a strong tradition of using peat as its main source of domestic fuel, but in November 1990, the city's peat store containing 76,000 tonnes of dry harvested peat accidentally self-ignited. On the front page of the 17th November 1990 edition of the *Uppsala Daily News*[24] is a black-and-white photograph of a huge pyramid of fire as cars pass close by in the dark during the early evening rush hour. It is accompanied by the headline '*Här går 100 miljoner upp i rök*' – 'Here goes 100 million Swedish Kroner up in smoke'. A thick cloud of smoke and the smell of peat reek engulfed the city for weeks, and it took two months to extinguish the fire completely. It is unclear how the huge 2019 fire that engulfed a large part of the peatlands of the Flow Country in Caithness and Sutherland actually started, but in just six days it burned deep down into peat and released 700,000 tonnes of carbon dioxide back into the atmosphere.[25]

Although natural peat fires and peat kilning are governed by the same principles, one is unmanaged and difficult to control, the other carefully managed. By controlling airflow, oxygen supply and fuel load the kilnsman is able to achieve the right balance between smouldering and flaring. If there is too much oxygen, smouldering rapidly turns to flaring, leading to less smoke and the conversion of most of the precious carbon-rich peat to carbon dioxide, water vapour and a mineral-rich residue of white ash. Nowadays, the maltster requires peat smoke, not peat heat, so once the fire is established the skill is to slow down airflow and starve the combustion process of oxygen – controlled pyrolysis. This results in surface smouldering rather than surface flaring and the release of gases and carbon-rich particulates, including large amounts of the important volatile phenolics, that contribute to sweet, smoky and peaty malt whisky.

Over the last 100 years, science has significantly improved our understanding of the combustion process and its by-products. Despite such progress, setting and controlling the fire in a peat kiln is a learned skill that has been passed down through generations of kilnsmen. Importantly, individual distilleries and maltings differ in the way they apply the craft and the science of fire, flame and smoulder, maintaining tradition and creating individuality in their peated whiskies. Science has not so much changed the way peat is combusted in the kiln to add flavour but revolutionised the understanding of the process, promoted experimentation and improved consistency. Research is currently underway to improve kilning efficiency and reduce peat use. To put it simply, if you can smell peat smoke in the vicinity of a kiln, important flavour compounds are escaping into the atmosphere rather than being adsorbed by the malted barley.

Kilning halts germination by drying the barley with a draft of warm air that dissipates through a grain-covered floor, a vast rotating drum or a box that is used to malt the barley. Most large-scale industrial maltings use fans and airflow controllers to dry the barley, although some distilleries such as Highland Park still use a traditional system that relies on natural draughts. The Doig Ventilator, a form of cupola shaped to resemble a pagoda roof, which is such a recognisable feature of many of Scotland's distilleries, was invented in 1889 and first installed at Dailuaine Distillery by Charles Cree Doig, the famous distillery architect, who created a system of kilning with a design that allowed air vents to be opened and closed to change the density of peat smoke.[26] In most cases, the drying cycle starts with a low temperature withering to make the barley hand-dry – this is the stage when the surface of the grain is most effective at adsorbing smoke. At Laphroaig Distillery, where they make a famous medicinal and phenolic Islay whisky, peat smoke is used for the first seventeen hours of the drying process, so the tarry phenolics can easily stick to the moist surface of the grain. After the film of water on the surface of the grain has disappeared, the temperature is increased to produce dry, crunchy malt that can be easily ground to a grist in the distillery mill.

Smoke can be managed in different ways in the kiln to control levels of phenolics. If the barley is too wet at the start of kilning, the thin film of water on the surface of the grain can prevent the phenolics sticking to its surface – if too dry, the smoke has nothing to promote its retention and adsorption. It is a highly skilled part of the whisky-making process – too much heat during kilning will denature and destroy the important enzymes that later convert starch into fermentable sugar. Over time, the traditional peat furnace has gradually become a secondary kiln to produce smoke, not heat. In the early days of distilling, the densest, most energy-rich bottom peat provided the heat to fire the stills and generate the energy for the kiln, while the fibrous top peat would be the best smoke generator. The emphasis now is on peat volatilisation, not peat combustion.

John Thomson has now retired to fly his glider in the skies above some of the distilleries and maltings of north-east Scotland, where he worked for forty-one years. During a long and varied career in the industry, he spent seven years as production manager at Port Ellen Maltings on Islay, where he made peated malt for the distilleries of Islay and Jura. We spent a pleasant Sunday morning on the Fife coast, talking about his career and the art and science of making peated malt. John is not entirely happy with his legacy in the whisky industry – quite soon after we started talking, he confessed, 'I was the man who knocked down Port Ellen Distillery.' I tried to soothe his conscience by saying that it didn't matter now as it was being rebuilt. I'm not sure it helped, but it was good to get that out of the way early.

In 2004, John wrote a thesis for his Master Maltster exam, *An Investigation of Factors Influencing Efficient Phenol Production in Highly Peated Distilling Malt*.[27] He studied many aspects of peating malt at Port Ellen, including malt quality and grain size, peat source and kilning time, and even tried to measure the temperature changes that went on inside the fire of the peat kiln: 'I borrowed a three-metre-long thermometer from the Scotch Whisky Research Institute laboratories to measure the temperature at the heart of the fire. It wasn't a great success; the temperature on the thermometer rose rapidly – 600ºC, 800ºC, 1200ºC and then pop! Diageo eventually coughed up for a replacement.'

Like most maltings and distilleries, Port Ellen smoulders its peat at the beginning of the drying process when the malt is moist. 'We used both the sorted, riddled peat nuggets and what in Islay they call "*caff*", a dross of peat fibres and dust. Our aim was to create a gradient of heat in the kiln, a white heat at the bottom, and a cold heat at the top.' Smoke is a visible and dynamic vapour produced during burning – above a peat fire, a vortex of suspended gases and flying particles containing aromatic smoky compounds rise steadily upwards towards the drying malt. Combustion in a low-oxygen environment breaks apart the large macro-molecules within the peat to create smaller compounds that are both volatile and odour-active. 'We are basically steam distilling and pyrolys-ing the peat by controlling the airflow and restricting flaring. A range of compounds produce the smoke character of malt, but the phenolics are relatively easy to measure and a good yardstick. The highest yield of phenols is in the heart of the peat fire where the kiln temperature is around 800°C.' Different kilning regimes are needed to produce malted barley with low, medium and high levels of phenols for the island's distilleries. 'Experienced operators know how to set and tend the peat kiln to get the right level of phenols in the malt and to create the optimal thermal gradient within the fire. It is still more of an art or a skill than a science.'

Some distilleries do it differently, creating individuality in their whisky. The kilning cycle at Highland Park Distillery on mainland Orkney starts with fourteen hours of peat smoke to add flavour to the malt which is then dried at a slightly higher temperature by burning coke for twelve to twenty-four hours. The peated barley is then mixed with unpeated malt to produce a less phenolic whisky, with a softer peat smoke flavour more akin to a lighter Highland style. Considering Orkney is one of the most peat-rich places in Scotland, this might come as a surprise. It has been attributed to James Grant from Glenlivet, who purchased and expanded the dis-tillery in 1895 and is credited with creating a whisky more akin to the lightly peated and fruity original Glenlivet from his home in Speyside.[28] Highland Park began to produce a style of whisky that was more attractive to the blenders. It turned out to be a smart move – Highland Park single malt is now one of the most impor-tant constituents of The Famous Grouse, Scotland's most popular blended whisky.

In his *Whisky Bible*, Jim Murray describes Ardbeg as 'unques-tionably the greatest distillery to be found on earth'.[29] Not everyone agrees, especially if peated whisky is not your thing, but along with its near neighbour Laphroaig, it is a name that is synonymous with heavily peated whisky made on Islay. In the 1970s, Ardbeg single malt was a key constituent of blends like Ballantine's and before it closed in 1981, produced a series of vintages that have become truly iconic.[30] The distillery was experimenting in its floor maltings with peating levels, different malt sources and mixtures varying from zero to 100% peated malt that resulted in the release of Kildalton, a famously unpeated Ardbeg expression. I have had the joy of tasting it: smooth, multi-layered, a creamy crème caramel, full of fruit with apple, citrus and pear and after a splash of water, extra spice, suet pastry and even my model plasticine from childhood – a stunning whisky and a demonstration of the quality of the spirit that the old Ardbegs were built on. At the other end of the spectrum, the old smoky Ardbegs were a powerhouse of briny, fruity, sweet and rich peaty flavours, and when the distillery closed, it was producing a highly phenolic malt that ex-master blender Robert Hicks eloquently described as 'the most pungent and powerful whisky in the world' that tasted like 'tarry rope or creosote'.[29]

When Ardbeg reopened in 1989, its floor maltings did not, and Port Ellen Maltings became its new malt supplier. The new owners of Ardbeg wanted their whisky to be more attractive to the blenders, and the large-scale centralised maltings just along the coast from the distillery were able to produce malt with a more consistent and lower phenol content. But still the modern Ardbegs, distilled after the distillery reopened, inspire whisky writers to new heights of peaty rapture. Dave Broom tasting a 1993 vintage, 25-year-old Ardbeg[31] described it as 'enormous, phenolic and filthy' – he detected hints of 'smoked duck, maybe even *Guga*'.[32] Olfactory hallucination maybe, but whisky tasting notes should be fun. I have also tasted a 1993 vintage Ardbeg – this time a 10-year-old – no avian hallucinogenics this time, just a sweetness and a smoothness, dry tobacco smoke and bitter Seville orange, but another object of great beauty.

In time, and as anthracite and coke became more widely available to the distillers through the expanding railway system and the coastal puffers, peat use in distillery maltings declined. Both these new fuels may have contributed to changes in whisky flavour, and although closed kilns became more common, coal dust and coke particles would still find their way into the drying malt above.[33] Even before coal and coke were added to the peat kiln, other materials were being used in the drying process. In 1698, Moray noted that heather faggots or blossoms were added to the peat to fire combustion and enhance aroma, although he said broom should be avoided due to its unpleasant aroma.[34] Small amounts of heather were once burnt in the kiln with peat at Highland Park, a practice that also took place for a time at Glen Ord and Glen Mhor distilleries.[35] In the same way that brewers add activated charcoal to their wash, it was at one time added to the wash still to clean raw spirit.[14]

In 1910, Glengoyne Distillery became one of the first, if not the first, to close its own floor maltings[36] and start buying in malt from other suppliers, but it was not until after World War II that the whisky industry started to change the way it malted its barley. Firstly, the Saladin Box arrived at distilleries like North British, Tamdhu and Dalmore, consisting of several metal forks that rotated as they moved up and down a trough filled with malting barley.[35] They were followed by drum maltings, a process of pneumatic malting that effectively pushes air through germinating barley, first patented in 1878.[33] At Port Ellen Maltings on Islay, seven vast rotating steel drums steep, germinate and dry up to fifty tonnes of barley over much shorter times than could ever be achieved by floor malting.[37] Both processes used hot air to dry the grain to make a consistent, lightly toasted and brittle malt. Peat, which had been the primary source of fuel, was now only used in secondary smoke generators, switching in and out of the malting process as required. One of the key benefits of centralisation of malting by the DCL in the 1960s was that it allowed production of malted barley peated to a more consistent level using a precise dual-fuel kilning cycle involving peat followed by anthracite. The company decided that all its mainland distilleries making peated whisky would now have the same level of peatiness – only Brora, for a while, and the island distilleries of Islay and Talisker were allowed to exceed this.[37] In the days before centralised laboratories brought more analytical precision to the production process, consistency was achieved by stipulating that the peat used at the maltings should have a moisture content of between 60–65% and be cut from the bogs in brick-like pieces with the precise dimensions of 5x10x25cm.[38] Consistency, homogeneity and repeatability became the core mantra for the big players in the industry, driven primarily by the global market for blended whisky.

Some experts still lament the change from floor to industrial maltings and suggest it has changed the nature of peated whisky, especially single malts made in the Scottish Highlands.[38] In the past, moisture levels in floor malted barley were kept higher and malting took a lot longer to complete as external factors like seasonal changes in air temperature made the process more unpredictable. It has been argued that the old floor maltings produced a less intensely flavoured malt with richness and complexity, resulting in a whisky with a lighter, softer, more mellow smoke. Peated barley produced in industrial scale maltings and largely destined for blends is regarded by some as harsher, more antiseptic and medicinal in character, exactly how many peated whisky fanatics like it.

Antiseptic Peat

Lullymore Bog in the Midlands of Ireland is famous for its antiseptic properties and preserved treasures. Since the 1600s, it has been drained, cut and mined for peat, turning it into an open, uniform industrialised landscape. Disturbance by humans has revealed some remarkable archaeological finds, including wooden trackways or *toghers* that formed part of a network of peat roads that connected communities across the wetlands of Ireland.[39] In the 1st century AD, this promoted trade and social interaction. The Lullymore Bog contains other preserved treasures that open windows to the past. It has yielded Bronze and Iron Age bodies that may have been ritual sacrifices to the Goddess of the Land. Ancient bog oak grew well here in the warm, wet climate of Ireland and has even found its way into the staves of modern-day whiskey casks. One of its most remarkable treasures is bog butter, a community offering to protect its valuable cattle – 4,000 years after burial, it is apparently still edible.

As well as butter, bodies and wood, the wet, antiseptic properties of peat have also been used to deliberately bury and hide whisky. Joseph Mitchell, a prolific engineer of Highland roads, bridges, harbours and canals, tells the story of the Laird of Borrodale, who, before joining forces with Prince Charles Edward Stuart in the 1745 Jacobite Rebellion, buried several kegs of whisky in a peat bog at Glenfinnan.[40] Unfortunately, the trustees to the whereabouts of the whisky were all killed in the Battle of Culloden and though they tried, the locals couldn't locate the kegs. In 1810, local peat cutters finally struck gold and the whisky that had undergone an unusual sixty- to seventy-year-long maturation period was unearthed and drunk. Mitchell was given some by the new laird and enjoyed it immensely, called it one of 'insinuating' quality. Just a few years ago, a stash of moonshine whiskey was found beneath the floorboards of a farm building undergoing demolition in Tyrone, Northern Ireland. About 200 bottles were revealed lying side by side, packed with peat turf and covered in insulating, antiseptic *Sphagnum* moss. Each bottle was handblown and likely dated from the period 1840–60. After more than 150 years, the corks had shrunk and the some of the contents had been lost, but other bottles had passed the test of time. One of the newly opened bottles had a nose described as 'bandagey, medicinal' and a finish of 'dried mud, burning turf'.[41]

The antiseptic properties of peat were well known to the medical profession. Peat could be slowly distilled to create a cleansing soap said to be an excellent treatment for skin diseases such as acne and eczema. The curative powers of a peat bath were at one time popular amongst the visitors to the spa towns of Strathpeffer in Scotland and Franzenbad in Bohemia. Not only the skin benefitted from a twenty-minute soak in 'moor earth': its powers of healing were said to include chronic rheumatism, sciatica, gout and nervous disorders. 'I felt a delight which continually increased at the enjoyable sensation of warmth as my body became more and more impregnated with the moor. All my nerves tingled,' said a glowing Miss Mary Fermor after an immersive experience in a black peat-bath.[42]

Lack of oxygen, darkness and low temperature all contribute to the preservative powers of peat. Importantly, it also contains high concentrations of organic acids that possess antiseptic properties that inhibit decomposers like bacteria. This property is well known and *Sphagnum* moss from peat bogs has been widely used to aid the healing of wounds. Records show that in 1014 at the Battle of Contarf in Ireland, it was applied as a wound dressing.[43] It was also used for the same purpose in the Napoleonic wars at the beginning of the nineteenth century. Its antiseptic properties, enhanced by a strong absorptive power, made it an attractive natural diaper for the infants of Native American women, wicking away urine and using the *Sphagnum's* acidity to avoid skin rashes. It was the world's first disposable nappy,[44] and the same group of women made use of its properties as a sanitary napkin. Dry *Sphagnum* is also a great insulator and Northern people lined their boots and mittens to keep out the biting cold. *Ötzi*, the Tyrolean Ice Man who reappeared in 1991 after 5,200 years entombed in an Alpine glacier, died with his boots filled with moss.[44]

Utilising its powers of absorption, healing and deodorisation, peat-wool and particularly *Sphagnum* dressings began to play an increasingly important role in times of war. This reached a peak in World War I as British army doctors at last recognised the special properties of the plant. *Sphagnum* acts like a sponge and is able to absorb twice as much blood or pus compared to cotton wool. Remarkably, it can hold more than twenty times its own dry weight of liquid, the main reason being that at any one time, only 10 per cent of the cellular plant structure is living.[43] The rest of the cells are dead and empty voids naturally fill with water or air. During wartime when cotton was in short supply and expensive, *Sphagnum* was collected from the bogs of Britain and Ireland by organised teams of women and children. By the end of the World War I in 1918, a million dressings were being made and sent to the theatres of war in Europe and North Africa.[45]

It is now known that the sterile acidic properties of *Sphagnum* are created by a simple piece of surface electrochemistry that involves charged particles called ions. Each *Sphagnum* plant cell is surrounded by a field of negatively charged particles[44] that attract ions, like calcium and magnesium, with an opposite positive charge to their surfaces. The larger incoming ions displace smaller and more weakly charged positive ions, such as hydrogen, from the cell surface, and this dynamic ion exchange process creates an acidic bog liquid with antiseptic properties.

At one time, distillers were even experimenting with the use of mosses in whisky making.[46] In 1887, Harald Breidahl, a Danish industrial chemist, arrived in Australia with his wife to take on the role of manager at the Federal Distillery in Melbourne.[47] Harald, who had previously worked for the Haig whisky dynasty in Britain, was also a keen bryologist – a collector and cataloguer of mosses. In his spare time, he scoured the peatlands of Victoria and amassed a considerable collection, some of which is now exhibited at the National Herbarium in Melbourne. In 1901, he filed a patent for the use of moss in whisky maturation.[47] What he was up to, no one is exactly sure as the patent cannot be found, but the most likely explanation is that he was using the moss as a 'natural purifier' to clean up the spirit during cask maturation. The ability of *Sphagnum* to soak up positively charged ions and trace metals could have been employed by Breidahl to improve the spirit quality in the cask. Alternatively, he could have been using the strong natural pigments of some mosses, particularly browns and reds, to add colour to the maturing spirit. Whatever he was experimenting with was not a commercial success and quickly fell out of use.

But what happens to the medicinal benefits of peat when it is burnt? It is not hard to find science that confirms the negative effects of smoke or smoking on health and in 1996, a research paper published in the medical journey *The Lancet*[48] put the Scotch whisky industry into a brief but angry tailspin. A study had showed that a carcinogenic group of compounds known as polycyclic aromatic hydrocarbons present in smoke were also present in variable amounts in different types of whisky. The researchers found the highest levels in Scotch malt whisky associated with peated malt or whisky that had been matured in charred casks. Despite concluding that the concentrations of polycyclic aromatic hydrocarbons in whiskies were low compared to smoked foods and were unlikely to be a cause of cancer, the study clearly unsettled the Scotch whisky industry.

But go back 150 years or even further and doctors and physicians were actively promoting the benefits of peat smoke, rather than being concerned about its detrimental effects. In 1685, the Archbishop of Dublin described a peat turf fire as the 'sweetest and wholesomest fire, that can be; fitter for a Chamber, and consumptive People, than either wood, stone-coal, or char-coal'.[49] A local story about whisky smugglers working close to the present-day location of the Aultmore Distillery relates that they once used the smoke of burning peat to preserve yeast for the next fermentation.[50] This was achieved by coating the cut branches and leaves of a birch tree with a yeast paste and hanging it out to dry inside a chimney above a smoking peat fire.

It was in the blackhouses and bothies of Ireland and the Scottish Highlands and Islands that the 'blessed peat reek' was protecting the population from a killer disease. In Lewis in the mid-1800s, local doctors were well aware of the apparent immunity of the islanders to tuberculosis and put it down to peat smoke – its prevalence in the blackhouse acted like a persistent fumigant and killed the tuberculosis bacterium. W. Anderson Smith in 1874 observed that 'tubercular consumption is seldom or never found among the natives who have remained in Lewis'. He went on …

> … our conviction is that we must look at the healthy effect of the blessed peat reek, with which, during half their existence, their lungs are impregnated. Whenever they leave the health-giving outer atmosphere it is to enter into a strongly antiseptic one.[51]

The Lewismen also smoked peat in times of hardship when they replaced tobacco with a peaty fibrous material called *calcas*. Apparently, it was 'hot and utterly vile, but it was there when cigarettes were not'.[52]

The crofters' precious cattle were also cohabitants of those bothies and a Dr Morgan colourfully recounted how the dwellings 'reek with the impure exhalations of these joint inmates … warmed by a peat fire kept constantly burning in the centre of the floor; the luxury of a chimney is often altogether unknown'. The inhabitants were generally free of the 'tubercular phthisis', something he put down to 'inhalation of the peat smoke and the antiseptic ingredients contained therein, the tar, the creosote, and the tannin, together with various volatile oils and resins'. When the residents left their blackhouses and went to the mainland, they often became infected with the killer disease.[53]

The beneficial effects of peat smoke were also well known in Ireland. In the 1870s, the 6,000 citizens of the town of Killarney lived in 'squalid cabins, furnished with the inevitable dung-pit'. But nevertheless, the town was a 'remarkedly healthy place' due to a good 'supply of the very best water and smoky houses'. The atmosphere rich in peat smoke acted as an 'antiseptic, deodoriser and preventative against infection and malaria'.[54]

Probably the most famous of the Lewis physicians was Dr Charles M. Macrae, who died at the ripe old age of ninety-one in 1909. 'Lewis's Own Grand Old Man', he studied arts, theology and medicine in Edinburgh, the latter under Sir James Young Simpson, who discovered and pioneered the anaesthetic use of chloroform in medicine. Macrae received a gold medal for his thesis in 1848 on the *Antiseptic Properties of Peat Smoke*.[52]

Medicinal Whisky

Alexander Fleming, the Nobel prize-winning doctor who discovered penicillin in 1928, had a favourite preventative for the common cold: 'a good gulp of hot whisky at bedtime – it's not very scientific, but it helps'.[55] Likewise, whisky was and still is used in hill farms in the Scottish Highlands to revive new-born 'cold' lambs in spring. Other 1950s remedies for the flu included hot rum punch, elderflower wine, a good hot breakfast, a raw onion, and if that didn't work you could always rub your chest with a blend of mustard and lard, or goose grease.[56] Winston Churchill, the wartime Prime Minister, like many politicians before and after him, was a practising advocate of the restorative powers of whisky, in his case, Johnnie Walker Black Label. Margaret Thatcher was famously never far away from her bottle of Bells.[57] If peat was the antiseptic preservative and peat smoke the preventative, whisky was both the preventative and the restorative – an all-purpose, cure-all, wonder medicine.

The perceived medicinal powers of whisky have long been attractive to the marketers, and official bottles of The Glendronach single malt bottled before World War II carried the slogan 'Most Suitable for Medicinal Purposes'. They were one of many – Johnnie Walker was also proclaiming its medicinal powers, and most famously of all was peaty, antiseptic Laphroaig. The known medicinal powers of whisky go back at least a thousand years and to this day many people, including myself, practice the mantra of 'little and often'. Before the mid-1600s when it started to be more widely used as an intoxicant, whisky or *uisge beatha* had been more commonly employed as a powerful medicinal agent. Distilling was a Celtic craft, and it may have been the Beatons,[58] an established family of physicians, who first brought the practice of distilling across the narrow stretch of water from Ireland to Islay in the 1300s. Although the list of its perceived curative and preventative powers was lengthy, it was in most cases used as a way of effectively relieving symptoms and promoting recovery by relaxation and sleep. You have to admire the circular logic of the makers of Irish *aqua vitae*, who used both peat and water from the bogs of Ireland for their craft, as a sixteenth-century ameliorative for the 'distillations, reumes and flixes' caused by the country's wet and boggy environment.[59] Antiseptic peat, antiseptic peat smoke and now medicinal whisky.

Martin Martin travelled extensively in the Western Isles of Scotland and recorded in 1703 how the natives of Harris drank *aqua vitae* or brandy in doses to combat colds. In Mull, one of the wettest of the west-coast islands, the natives drank *aqua vitae* as a 'corrective' to counter the dampness of their climate. In North Uist, it was used to treat many ailments including diarrhoea – 'the stronger the better' – and in Skye, the locals drank three times more strong liquors than in southern Scotland to protect them from the 'moist and cold' climate of the island.[60]

Whisky appears regularly in the writings of those early travellers, engineers and geologists who roamed the Highlands of Scotland, making maps and building roads and bridges. It was very much part of the routine of daily life and widely available to paying guests and given as a welcome or a parting *Deoch an Doras* to unexpected visitors. Lodging at a house in the island of Skye in the 1800s, the geologist Sir Archibald Geikie recounts a tale of the first whisky of the day, a common way of heightening the appetite before breakfast.[61] It was a meal he clearly looked forward to, writing 'there are few meals in the world more enjoyable than a true Highland breakfast'. The eldest daughter of the house brought a 'tray laden with bottles and glasses' to his bedside and an offer to 'taste something before I got up'. After initially refusing she offered again, 'some whusky nate? some whusky and wahtter? some whusky and milk?' He consented to 'whusky and milk' before launching into what he termed a 'sensible substratum' of porridge and cream followed by the rest of the breakfast table, before heading out into the cold and rain to make his maps and collect his rocks.

In the growing industrial towns and cities of Britain, spirits and beer had an important role to play in the prevention of disease caused by poor sanitation and bad water. One of the reasons the British working class became great tea drinkers was the importance attached to boiling water as a way of combating waterborne diseases such as typhoid and cholera. In the same way beer, a fermented alcoholic beverage, afforded a higher level of protection from disease than drinking unboiled water. Spirits, in particular gin and to a lesser extent whisky, would have also been viewed by those who could afford it as a safer alternative to unboiled water.

In the nineteenth and first part of the twentieth century, it was common to see whisky endorsed with statements proclaiming its medicinal benefits. Johnnie Walker ran advertisements in the 1910s using testaments from the medical profession to promote its powers. In 1940, Old Bushmills was being advertised to 'Kill a Cold' and in the 1950s, a Jameson advert even featured a white-coated doctor using his stethoscope to examine a bottle of 'Superior' Irish whiskey.[62,63] Although it is not clear when whisky began to fall out of favour with doctors, at least officially, in the inter-war years Brompton Cocktail became a popular alcoholic 'elixir' widely used by the medical profession.[64] It was made up by a pharmacist using gin, whisky or brandy with specific amounts of morphine and cocaine and was prescribed for pain relief, particularly amongst the terminally ill. The mixture was used until the 1980s and early 1990s when it was replaced by better directed modern drugs.

Although there is an old adage from the 1960s that says, 'an alcoholic is someone who drinks more than his doctor', there is little doubt that many in the medical profession at one time believed that whisky, administered in small amounts and with regular doses, could have beneficial effects on patients. In 1903, the *British Medical Journal* published an article simply called *Whisky*.[65] It reflected upon the rise in consumption of whisky in the population and invited its practitioners to accept a degree of responsibility for its growing popularity. What then followed was a discourse on early whisky chemistry, the imperfections of current analysis, and the lack of understanding of the physiological actions of the individual constituents.

Lt Col Stuart Henderson Hastie, OBE, MC, BSc, FRIC

Stuart Henderson Hastie has been called 'probably the first chemist to be employed in malt distilling'.[66] He was born in 1889 in a tall, sandstone townhouse in a handsome part of Edinburgh,[67] just across the road from where I first appeared in the world. In his twenties, he made an unconventional start to a career as a whisky chemist by becoming a tank commander. He went to war in the battlefields of Northern Europe, naming his own killing machine 'Dinna-ken', and was awarded a Military Cross after the Battle of the Somme.[68] In the post-war years, he left military service to become a 'practical chemist' in a local Edinburgh brewery, before Sir 'Restless' Peter Mackie of White Horse Distillers fame decided to invest in science by employing him to set up a new laboratory at the Hazelburn Distillery in Campbeltown.

Hastie's time at Hazelburn was preceded by Masataka Taketsuru, the 'Father' of Japanese whisky and founder of the Nikka Whisky Distilling Company.[69] Taketsuru was already a qualified chemist when he arrived in Glasgow in December 1918 and after studying organic chemistry at the university, spent 1919 visiting distilleries to learn and record in detail the art of making Scotch whisky. He went on the train to Speyside and turned up at Glen Grant, Glenrothes and Longmorn, before travelling to Campbeltown at the beginning of 1920 with his newly-wed Scottish wife Rita. This became the most important part of his stay in Scotland when at the invitation of Peter Mackie, he spent four key months at Hazelburn Distillery. Taketsuru was a scientist and his rigorous, experimental and enquiring mind would have rubbed off on all who met him, including the distillery manager Peter Innes and Sir Peter Mackie. After his time at Hazelburn, he felt he had learnt enough about Scotch whisky and returned to Japan to set up the Yoichi Distillery on the peat-rich island of Hokkaido. He wanted to make a Highland-style, heavy, peated spirit.

It is unlikely that Henderson and Taketsuru ever met, but it cannot be coincidental that soon after he left Scotland and returned to Japan, Mackie, realising the growing importance of science and analysis in the whisky industry, set up the chemistry laboratory at Hazelburn and employed Hastie to run it. This is often hailed as the first distillery laboratory, but Barnard noted a 'small laboratory for experiments, and a library of scientific books' when he visited Yoker Distillery near Glasgow almost forty years earlier. Hastie's chemistry, which is recorded in copious amounts of paper records, supported Mackie's production at Hazelburn, Craigellachie, Lagavulin and the legendary Malt Mill Distillery. When Hazelburn Distillery closed in 1925, such was the importance of Mackie's laboratory and its analytical records to the company that he sent Hastie to reclaim all the laboratory equipment and data and bring it back to head office in Glasgow.

Like his boss Mackie, Hastie was a colourful and outgoing character, unafraid of ruffling establishment feathers. On 8th February 1926, aided by a few glass lantern slides, he read a paper to the Institute of Brewing at the Engineers' Club in London entitled *Character in Pot Still Whisky* that became something of a benchmark for the industry.[14] Hastie, who was the first chemist to understand the role and importance of the colourless oily and aromatic liquid furfural in whisky flavour, used the evening to vent his frustrations on those in the room. He was frustrated with the distillers and frustrated that the tools at his disposal as a chemist were inadequate to answer the questions of his time. He recognised the importance of local peat and water quality in whisky making, and knew that a link existed between the two. He also recognised that different sources of peat from different parts of Scotland had an influence on the character of local whisky. But as a chemist, he had no evidence to prove that these links existed, saying 'chemical analysis as at present employed is useless for these purposes'. He knew that most distilleries still lagged behind the brewers and powerfully drove his point home, concluding 'distilleries were suffering, and suffering badly, from the absence of the application of scientific methods'. He hoped it would be a loud wakeup call to the industry.

While at the end of the nineteenth century some companies like William Teacher's & Sons were already paying public analysts and laboratories to perform routine chemical analysis on samples of whisky and write effusive recommendations regarding their 'excellent quality and fine flavour', fifty years before Hastie harangued his audience in London, DCL was only employing chemists on a short-term basis. This was largely to sort out waste disposal problems or to find out what their competitors were up to.[70] In 1896, it eventually employed its first permanent chemist and started using the new analytical techniques that were emerging in the late 1800s. These would go on to revolutionise the industry in fields like production, quality control, waste management, malting, flavour analysis, chemistry of wood maturation and, most importantly, blending.

Hastie, who died in 1980 in Elgin, worked for White Horse Distillers and SMD before becoming a director of DCL in 1948. His great legacy was a sixteen-page booklet *From Burn to Bottle*,[71] a detailed and very precise step-by-step guide on 'How Scotch Whisky is Made'. Dr Nick Morgan, former Head of Outreach at Diageo, called it 'very much Hastie's manifesto for the future of the industry'. It reads like an old-fashioned recipe book and leaves the reader in no doubt as to how he felt Scotch whisky should be made. Distinctly old-school and somewhat arrogant, Hastie continued to ruffle feathers and in the booklet he wrote:

It is a fact, however, that Scotch Whisky cannot be made anywhere else. The Japanese came to this country years ago, copied our plant, and even employed some of our Speyside personnel. They produced an imitation of Speyside Whisky which was not good, although drinkable.

Faint praise indeed. Hastie may have been hailed as a pioneering whisky scientist, but ironically, it was a young chemist from Japan who appears to have been instrumental in creating his job in the first place.

In many ways, Hastie was ahead of his time and he began working in distilling when chemistry was only slowly beginning to make its mark on the whisky industry. Traditional practices and old prejudices handed down from operator to operator were now being challenged and changed by scientific methods. Morgan is clear about the debt the industry owns to Hastie.

Stuart Henderson Hastie was a hugely influential figure in DCL and his philosophy, based on scientific analysis and testing, set a benchmark that is still hugely important to this day. You can bet your bottom dollar that before a change in production process occurred, like a switch from direct firing of stills to steam distillation, it would be tried, tested and tested again. Diageo's philosophy is very interventionist and that is all about consistency – a philosophy that started with Hastie at DCL. If a problem arose, 'The Men in White Coats' would be sent in.

From 0 to 309 ppm Phenols

At the beginning of the twentieth century, two American industrial chemists made a significant contribution to unlocking the smoky gene when they detected traces of 'pyrol and phenolic bodies' in whisky.[1,16] One of them was Philip Schidrowitz, a German-born industrial chemist of some repute and an expert on spirits who in 1908 gave evidence before the Royal Commission on Whisky. One year earlier, he had presented a paper to the Institute of Brewing in a meeting room at Queen Street Station Hotel in Glasgow. He took the opportunity to both congratulate the brewers for embracing the new science, and like Hastie almost twenty years later, lambast the majority of distillers in the audience for being stuck in the past. He illustrated their conservatism by using a story about a Highland distiller who refused to replace a manual hoist with a mechanical device for lifting grain as it might affect the quality of the whisky. True or not, the point was well made. He foresaw a major future role for chemists in the art of blending and was 'firmly convinced that the day will come when the blender and the chemist will work hand in hand, and that art and science combined will achieve results beyond the reach of either'. He turned out to be right.

While Hastie was a strong and vocal advocate of the role of chemistry in the industry, he wasn't the first whisky scientist and was preceded by well-known figures like Schidrowitz and J. A. Nettleton, the latter a highly regarded industry chemist. However, the first person with real power and influence to understand the role that science might play in future whisky production was Henry Ross. Son of William H. Ross, the chairman of DCL and the man credited with creating the basis for the success of the modern-day Scotch whisky industry, Henry learnt his chemistry at Heriot Watt College in Edinburgh and later in Germany, where he saw for himself how scientific knowledge and experimentation underpinned its world-leading chemical industry.[72] He worked in the company's distilleries and in overseas marketing before becoming the youngest member of the DCL Board in 1925. His task was to set up and promote research and technological innovation within the company. Two years later, he purchased a country mansion on Epsom Downs in southern England and turned it into the company's new technical and research department with modern laboratories and an experimental plant. The whisky industry was at last heading in a new and better direction – it was future-proofing itself with science.

In a climate of growing research and development in the industry, an analytical method arrived in the late 1960s that was to transform the making of peated whisky.[73] It was developed by Colin Macfarlane, a university-trained chemist who worked for Munton & Fison Ltd, an English company with a long tradition of making and trading malt both at home and abroad. In the 1960s, their malt trade with the Scotch whisky industry was beginning to grow and in 1963, the company entered into a partnership with Invergordon Distillers and Highland Distilleries to provide malted barley, some of it peated, for their own distilleries, which included Highland Park, Bunnahabhain and Bruichladdich.[74] It also made the popular Black Bottle and Famous Grouse blends that required a peated component. The Macfarlane Method involved extracting the adsorbed phenols from the surface of malted barley, oxidising the free phenolics and complexing them with a dye called 4-aminophenazone. The more intense the colour, the higher the phenol concentration of the peated malt. In the course of his research, Macfarlane visited Islay and met Bessie Williamson, who furnished him with samples of peat, peaty water and some Laphroaig single malt whisky to take back to England. In his laboratory, he confirmed that the unique character of Islay malt whisky was indeed caused by unusually high levels of smoky phenolic compounds, produced by the burning of the island's peat. He found an absence of phenols in the Islay water and concluded that it made 'no direct contribution to the peaty character'. Williamson wasn't happy, still convinced the special peaty Islay water contained phenols that enhanced the 'essence' of her malt whisky,[75] a nostalgic view still shared by some to this day.

Since the method first appeared in 1968, it has evolved to become more accurate and reproducible, and is still the industry's standard for measuring 'total phenol' concentration in light (1–5), medium (5–15), heavy (15–50) and super heavy (greater than 50ppm phenols) peated malt. Today in the small laboratory at Port Ellen Maltings, the operators steam distil the phenols off samples of malted barley and then use a modified Macfarlane Method with slightly different reagents, but with basically the same underlying chemical principles and colorimetric analysis.

The origins of the method in a malting company goes some way to explaining why the industry still uses the rather odd convention of expressing the phenol content of their whisky using the in-grain ppm phenol concentration of the malted barley. In the malting industry, the measurement makes sense and facilitates the trading of peated malt at home and abroad. But it is hard to think of another product in the food and drinks industry that states the starting concentration of a constituent rather than the residual concentration in the final product at point of sale. Open to misunderstanding, it is one of the odd and quirky things about the whisky industry that makes it so interesting, or frustrating, depending on your point of view. Several companies have tried to buck the trend and express the ppm phenol concentration in the whisky at the point of bottling. Knockdhu Distillery, producer of AnCnoc single malt whisky gave up on the idea, while William Grant & Sons Ltd and the new Torabhaig Distillery on Syke still fly the flag for the consumer. The whisky industry has been unwilling to embrace change despite the cost-effectiveness and speed of modern analysis.

The industry convention is even more difficult to comprehend when as much as 60–80% of the malty phenolics can be lost during the production of new-make spirit.[76] After that, the only way to increase the ppm phenol concentration is to fill the spirit into casks that previously contained heavily peated whisky – a way of chemically retro-fitting the spirit with smoke. Conspiracists argue that the brand marketeers like to keep the ppm phenol number as high as possible to help sell their product. Losses occur at different stages of whisky making, but one of the largest single losses occurs during distillation, and the point when the middle cut is made is crucial. In the words of Robert Hicks, 'The peat in Ardbeg comes through at the tail end of the main run' and we 'alter the length of the run to make sure we get the peatiness, but not the nasty stuff'.[30] Phenols appear late in the spirit run, so delaying the cut until around 60% abv increases their total concentration in the final spirit.

Back on the Fife coast, I asked John Thomson when did chemistry and science really begin to start to make a big difference in whisky making? 'Chemists began to play a leading role in the industry in the late 1950s and 60s, and by the late 70s and early 80s every distillery had opened a small laboratory for testing and analysis. Some of the companies had larger centralised laboratories.' John started his career in the whisky industry as a technician in a small distillery laboratory. He is a lad from Dufftown who left school in 1976 at the age of eighteen. 'It all began when I skipped double French. I didn't like languages.' He stayed behind after assembly to listen to a man from J&B who was recruiting for a laboratory technician to work in the company. 'On 1st August 1976, I went to work at Strathmill Distillery in Keith, learning on the job how to make routine tests of spirit, with day-release to college to study the chemistry behind the trade.' Their big seller at the time was the J&B Rare blend with Knockando, Auchroisk, Glen Spey and Strathmill single malt whisky at its heart. John took every opportunity that came along to learn new skills and get qualifications, and soon moved along the road to work in the main company laboratories in Elgin.

In the 1960s, the gas chromatograph (GC) arrived on the scene and led to the identification of more than 100 aromatic flavour compounds in whisky. The method involved the chemical separation of distinct groups of high alcohols in whisky and brandy,[77] and GC was one of a group of new instruments based on chromatography that, more than anything that had come before, started to unlock the smoky gene. The principle of chromatography is simple and involves the separation of the components of a mixture that are carried in a moving or mobile phase, either a gas or a liquid, by passing them over a stationary phase. In the case of phenols, the volatile flavour compounds produced by combustion each have a unique chemical structure that makes them more or less attractive to the stationary phase. This means they can be separated by chromatography and their peak concentrations measured using a suitable detector. The best analogy to explain the principle of chromatography is that of a mixed cloud of flying insects that is being gently blown across a field of flowers in a stream of air. The butterflies are highly attracted to the nectar of the open flowers and take longest to arrive at the edge of the field; the bees linger longer but travel faster than the butterflies; and quickest of all are the wasps who completely ignore the flowers in search of sweet rotting apples in a distant orchard.

The 1960s were a period of major progress in the world of whisky chemistry and a suite of techniques arrived one after another that all had one thing in common: they were capable of separating and detecting individual compounds. Gas chromatography (GC), high-pressure liquid chromatography (HPLC), gas chromatography mass spectrometry (GCMS) and ion exchange chromatography (IEC) were in terms of speed of analysis, sensitivity and selectivity, miles ahead of the old laboratory methods. A modern whisky analytical laboratory is now heavily reliant on GC and HPLC to cover the full spectrum of active flavour compounds in whisky, the first to measure volatile components, the second to measure liquids that are less easily volatised. By the early 1980s, the number of detectable constituents of whisky had grown to 300; by the early 1990s, it had reached 600.[78] Now the number is in the 1000s. Fifty years on from the time when Schidrowitz and Hastie vented their frustrations to audiences in smoke-filled rooms, the whisky industry had finally embraced the tools that they craved and was steadily beginning to unlock the fundamentals of whisky flavour and smell.

By 1973, Macfarlane had moved on from successfully developing a method to measure total phenols and started to pick apart the constituents and volatile flavour compounds contained within peat smoke, using the new instruments at his disposal.[79] As the ink dried on the chromatographs slowly rolling off the printer attached to his GC, he first observed a peak of furfural, followed by guaiacol, phenol, p-cresol, 2,3-xylenol and finally a small peak of 3,5-xylenol. This was just the beginning and by the 1980s, HPLC, backed up by GC, had become the most established method for detecting individual flavour compounds in whisky. Compared to the Macfarlane colorimetric method, the chromatographic methods were more accurate and precise and produced higher absolute values since they were capable of detecting all the important volatile phenols present in peated whisky, rather than just a few. Despite this, both the standard Macfarlane Method and the more precise chromatographic methods are still widely used to this day. Why there is no agreed standard approach is a mystery to a scientist and just another peculiarity of the whisky industry.

No one can doubt, however, that the Macfarlane Method triggered the start of a growing interest in peated whisky that for some became a passion and for others an obsession. In 2008, Bruichladdich Distillery on Islay released 6,000 bottles of Octomore 01.1, a 5-year-old single malt whisky bottled at 63.5% and on the bottle were the words 'using barley malted to a whopping 131ppm phenols'. It had taken the whisky industry forty years to start using compound-specific analysis as a marketing tool, and at the time it was heralded as the world's most heavily peated whisky. By 2014, peating levels in Octomore 06.3 had reached 258ppm and in 2018, the distillery released Octomore 08.3, made from malted barley with a staggering phenol content of 309.1ppm. The distillery called it 'unprecedented', 'a thunderstorm of a dram' and 'unlike anything that anyone has tasted before'. Malcolm Bruce, the distillery manager at Knockdhu, once told me that Octomore was so phenolic that you could 'stand your own pen up in it'.

Inspired by ppm phenols, the marketing of Octomore started a phenolic deluge of new whiskies with names like Big Peat, Smoky Scot, Smokehead High Voltage, Iron Smoke and Smoky Goat, apparently inspired by the foul-smelling wild goats that roam unmolested along the sandy beaches and cliffs of Islay's south coast. It had become not just an important analytical measurement for the whisky makers, but a need-to-know parameter amongst consumers.

Cresol, Guaiacol, Phenol

In front of me is a collection of small glass bottles purchased from a laboratory supplier, filled with cotton wool. Each one is labelled with the names of a chemical compound. There are eight in total and Dr Barry Harrison, the whisky industry's leading peat smoke chemist, invites me to nose each one in turn and put into words what I smell. This is how a sensory panel is trained; it feels like a test and I want to do well. These are my notes:

o-Cresol – soft, subtle, antiseptic bandages.

p-Cresol – stronger, antiseptic and powerful, a first-aid box.

m-Cresol – an overwhelming odour of antiseptic bandages, nothing else.

4-Ethyl Guaiacol – medicinal, spicy, sweet cloves – I'm at the dentist.

4-Methyl Guaiacol – sweet, cheap perfume, fags and vanilla.

Guaiacol – smoky unburnt ash, a smouldering Swan Vestas match that fails to light.

4-Ethyl Phenol – soft and medicinal.

Phenol – soft and medicinal, sweet and mildly spicy, with a hint of pear drops.

The bottles fall into three groups or 'phenolic families', each with a particular nose: phenols – soft, sweet and medicinal; cresols – stronger, medicinal with a powerful smell of antiseptic bandages and first-aid box; guaiacols – less medicinal with sweetness and spice. What is astonishing is that in a group of compounds like the cresols, the repositioning of a single methyl group on the core aromatic ring can have such a profound effect on aroma.

I am inside the Scotch Whisky Research Institute (SWRI) just outside Edinburgh, a neat modern building with a trademark pagoda on the roof. Formally Pentlands' Scotch Whisky Research Ltd and set up in 1974 by a group of seven independent distillers, it now forms the research arm of a much larger grouping covering most of the industry. A local curiosity in a modern research park, the pagoda acts like a beacon for new visitors and a sure sign that whisky lies behind the walls. We tour the research laboratories and Barry answers my many questions. The cereals laboratory with a mini-maltings and Petri dishes full of Optic, Concerto, rye, maize, millet waiting to be germinated. In the darkness of a room, a long-term cask maturation experiment under strictly controlled conditions of light, temperature and humidity. A sample preparation laboratory containing benches full of whisky bottles, some recognisable, others in labelled sample bottles – never have I seen such a fine array of liquids before in a laboratory. In a fume cupboard, one of Barry's essential pieces of kit, a cylindrical peat combustion furnace with a smoke collection system. Next door, at the heart of the SWRI, is a suite of top-of-the-range sparkling white analytical instruments: mass spectrometers with their particle accelerators and automatic delivery systems, HPLC, and a GC Olfractometer (GCO) – a GC with a big black flexible tube leading to a collection box with a face mask, an instrument that combines quantitative analysis with the real nose of a person. GCO separates the whisky into its individual components and then a seated operator identifies which ones are odour-active. The threshold concentrations of many of these aromatic compounds are often inconceivably low, at the ppm level or even lower. Whether they can be detected by us or not often depends on what they are dissolved in, be it water, wine or whisky. This is a room where modern analytical chemistry meets the whisky industry head on and shows just how far the science of distillation and whisky making has travelled in the last 100 years, and where it might be heading in the future. Some of the high-tech instruments in this room are even used to fingerprint and check the authenticity of old and rare whisky.

We then walk into the sensory laboratory with its six identical cubicles, each with a computer separated by a screen. Hanging upside down on the wall are three rows of uniform white and blue tasting glasses, and taking centre stage is the famous SWRI Flavour Wheel, an illuminated, multi-coloured dartboard of tastes. I pick out one of the core sectors with the word 'peaty'. This subdivides into burnt, smoky, medicinal and at the outer margins the words tar, char, ash, burning wood, smoked foods, BBQ, bonfire, antiseptic, TCP, carbolic soap and first-aid kit.

I asked Barry whether he thought peated whisky had changed since the 1960s with the arrival of industrial-scale maltings. He felt that although it was unlikely scaling up the malting process would have affected flavour, if the peat source or combustion temperature had changed then there could have been an influence on whisky taste. Until then, distilleries largely relied on local peat bogs to provide the flavours and fuel for malting, but as maltings became centralised, so did peat source. Phenolic compounds are just one of three types of flavour compounds produced when peat is burnt; the others are non-phenolics such as the furfurals, associated with an almond-like or sweet biscuity flavour, and the hydrocarbons, a catch-all term of under-researched complex compounds. They all contribute to the important group of compounds known as con-geners that give whisky its taste and smell – all the rest is water and ethanol. Some experts believe they can identify two types of peatiness: the first an Islay-style produced by pyrolysis and steam distillation in an oxygen-starved kiln that results in a heavy, peated whisky high in phenols with strong medicinal flavours. This differs from a more traditional Highland-style produced from kilning at higher temperatures in a more oxygen-rich atmosphere that results in a malt lower in phenols, but with more guaiacol and furfural. Peat combustion temperature and oxygen levels in the kiln con-trol the relative proportion of the different phenols. Many modern blends rely on Islay-style heavily peated whisky, produced in large industrial maltings, to provide that important heavy flavour com-ponent. In the past, the more lightly peated, softer Highland-style was an important component of many popular blends.[38]

Angus MacRaild is convinced that the flavour of peated Scotch whisky changed when the industry switched from floor to indus-trial maltings.

I don't think there were any distilleries that were making peated whisky that didn't experience a shift in flavour profile, although Ardbeg or Highland Park did retain some of their idiosyncratic peat character in later production years. Older style peat flavour can be characterised as richer, earthier and more herbal with more medicinal qualities. They tend to be more organoleptically complex. Modern peat profiles are drier, quite sharp, ashy, saline and citric, and lack the depth and more 'organic' characteristics of the older style examples.

The guaiacols, cresols and phenols that I smelt in the small glass bottles are all types of phenol – the group of compounds built around a single aromatic ring of six carbon atoms. Unlike the larger macro-molecules of the polyphenol group, they are small, volatile and in many cases odour-active. Guaiacol, which has a boiling point of 205°C, was first isolated chemically in 1826 by a twenty-year-old German chemist called Otto Unverdorben.[80] It is present in some of the traditional components that make up a good Scottish breakfast – a bacon roll, a kipper, a strong cup of roasted black coffee and a smoky dram of whisky. Add to this some cough syrup and the guaiacol feast is complete. Phenolic compounds as a group have been used in medicines and tonics for centuries. More than 100 years ago, an *Advisory to the People of Brazil* (*Conselhos Ao Povo*) from the Inspector of Hygiene about the Spanish Flu (*A Influenza Hespanola*) listed a range of treatments including 'taking water with iodine, citric acid and tannin as well as infusions containing tannin, such as those from guava leaves'.[81] Tannin is a light-brown phenolic compound and guava leaves contain many substances with known beneficial medicinal properties, including guaiacol.

The original black tea, lapsang souchong from the mountains of China, gets its smoke from the tar flavours and guaiacol produced by burning pine wood.[82] Guaiacol can be synthesised in the laboratory from the plant extract catechol derived from the leaves of the acacia tree. It is a simple chemical compound, often known as 2-methoxyphenol, a phenol with a methyl group right next to the hydroxyl group at the top of the aromatic ring. In the natural world, it is found in various plants and insects including locusts, where it is synthesised by gut bacteria that play a key role in the decomposition of vegetation. The locust's guaiacol is then used to produce an odour-active pheromone, a hormone that triggers a social response in members of the same species. The locusts are attracted to each other, gather in large groups and swarms, and often wreak havoc on local croplands.

Exactly eighty years after Stuart Henderson Hastie addressed his audience in the Engineers' Club in London, Barry Harrison published the first of several studies that showed conclusively that peats from different parts of Scotland were chemically distinct.[83] He was in the final stages of his PhD at Heriot Watt University. A few years later and now based at the SWRI, he followed this up with a paper that used an instrument that combusted peats at low temperature and identified the phenolic compounds in pyrolyzed peat samples.[84] He not only found differences in the amounts of guaiacols and phenols produced by pyrolysis of Islay, Orkney and mainland peats, but also differences in peat samples collected from different depths within the moss. Each peat source therefore carried a characteristic chemical signature and some, like peat sourced from Orkney, contained a distinctive suite of phenols, cresols and guaiacols that would help unlock one of the most interesting whisky stories of modern times.

In 1907, Sir Ernest Shackleton, a renowned teetotaller, took twenty-five cases (300 bottles) of Mackinlay's 10-year-old Rare Highland Malt Whisky south to the Antarctic on the *Nimrod* expedition.[85] The aim was to reach the South Pole, and they almost did, but on 9th January 1909 at 88°23'S, they were forced to turn back just ninety-seven miles short of their goal. Shackleton, Adams, Wild and Marshall had got closer to the pole than any other human beings, but in Shackleton's words, 'we have shot our bolt'. For them, survival was now all that mattered. Fifty days later, they staggered into the empty Discovery Hut and were rescued the next day. After eighteen months in the Antarctic, they left the expedition hut at Cape Royds in March 1909. With winter quickly arriving, his ship *Nimrod* departed the Antarctic bound for New Zealand with Shackleton and his team onboard. Three remaining cases of Mackinlay's were left under the hut and soon became a whisky frozen in time.

One hundred years later, as the ice began to thaw beneath the hut, bottles of Shackleton's forgotten whisky began to reappear and the whisky world stood up and took note. After thawing in New Zealand in 2010, three bottles were flown back home to the Invergordon spirit laboratory, then opened, tasted and ana-lysed. Never before had a whisky produced in the 1890s been so fully characterised using the full weight of modern analytical chemical and sensory analysis.[86] Measurements included acidity, alcohol content, carbon age dating, sensory analysis, ionic con-centrations, congeners and peat-derived phenolics. The results surprised everyone, not least master blender Richard Paterson ('The Nose') whose job it was to recreate Shackleton's whisky for twenty-first–century drinkers. He chose twenty-five different whiskies with ages ranging from eight to thirty years for the task. Those involved described the contents of the 100-year-old whisky as modern in style, 'a surprisingly light, complex whisky, with a lower phenolic content than expected'. Lightly peated, the whisky contained phenol, guaiacol and furfural in concentrations that matched the profile of the traditional lightly peated Highland style. Not only that, but a fingerprint of the types of peat-derived phenolic congeners matched exactly that of peat sourced from the islands of Orkney.[84] At the time, Mackinlay's were produc-ing high quality malt whisky at their Glen Mhor Distillery in Inverness and records show that they were shipping their peat from Eday. Sir Ernest Shackleton's choice of whisky was not that of a teetotaller, but of someone who apparently knew a good peated malt and valued its restorative and preservative powers for his men over the long Antarctic days and nights. The secrets of a 100-year-old malt whisky had been unlocked by modern analytical chemistry.

Two editions of the replica whisky were created by Paterson and the second (The Journey) contains single malts that include a 1980 vintage Glen Mhor and a rare heavily peated Dalmore. Bottled at 40% abv, his recreation of the Shackleton whisky has not unexpectedly divided opinion: 'I can't believe the great explorer Shackleton, would have taken such a rubbish whisky on his expedition', 'Very smooth, no burn, really mellow and very easy drinking and I love it' and my favourite, 'If I was in the middle of the Antarctic in a tent freezing my nuts off like the great man himself, this would be a great comfort'.[87] Personally, I was underwhelmed by The Journey, a smooth, fruity and malty easy drinker. I wanted to detect peat smoke but couldn't. Unlike Shackleton, I am not a teetotaller, but I for one would be wanting to take something a little more fortifying to the Antarctic.

'A distinctly different whisky.'

Original Cask Strength 10-year-old Laphroaig is billed as the ultimate challenge for the serious whisky connoisseur. Bottled straight from oak barrels, in which the whisky has matured for ten years on the Island of Islay, its peaty and uncompromising taste has earned Laphroaig the reputation as one of the world's most distinctive drinks. For those not brave enough to try a whole bottle, Laphroaig is also offering 'starter kits': a miniature bottle and a branded shot glass – the perfect stocking filler. The favourite whisky of the Prince of Wales, Laphroaig is the only single malt to be awarded his Royal Warrant. Available from all good off-licences.[88]

Laphroaig is famous for its advertising campaigns, challenging the drinker to be brave with slogans like 'no half measures', the 'ultimate challenge' and 'you'll always remember your first Laphroaig'. It is to lovers of peated whisky what a hot, sour and spicy vindaloo is to curry fanatics. Even during the time of Queen Victoria, it had a reputation as a heavily peated whisky, made by slow kilning over a smoky Islay peat fire to produce unusually high levels of phenols in the malt. Cold-smoke kilning produced the phenol-rich, medicinal properties that have become the stuff of legend.

Whisky folklore tells us that Laphroaig was the only Scotch single malt whisky that continued to be exported to the USA during the dark years of Prohibition (1920–1933), where it was prescribed as a medicinally flavoured alcoholic spirit that could be legally sold as 'medicinal spirit'. The story goes that Ian Hunter, the distillery owner at the time, used a loophole in American law that allowed whisky to be imported for that purpose.[75] Laphroaig rarely interested US Customs as they thought something that smelt like disinfectant couldn't possibly be attractive to drink. Laphroaig was noted for its iodine-like character and smelt strongly of antiseptic bandages and TCP. When the Volstead Act was repealed in 1933, the news that America had gone wet was celebrated in whisky-making towns and islands across the length and breadth of Scotland.[68] Nowhere more so than in Islay, where on 28th June 1934, Ian Hunter was quick to apply for the Laphroaig trademark to be registered in the USA.[75] He had done his homework well; his medicinal whisky was now in pole position to exploit the newly opened US market.

Before the start of Prohibition in the US, whisky and brandy were being widely used by doctors as a cheap and common treatment for a variety of ailments including anaemia, high blood pressure, heart disease, typhoid, pneumonia and tuberculosis. In other countries like Norway in the 1920s, whisky could only be legally purchased with a doctor's prescription.[70] In much of

Canada, a wholesale drugs licence was required. Whisky was prescribed to patients of all ages; a common dose for an adult was one ounce every two to three hours, for a child up to two teaspoons every three hours.[89] The drugstore Walgreens flourished during Prohibition and prescriptions could be exchanged every ten days for one-pint 'For Medicinal Use Only' bottles.

In the 1920s, iodine deficiency in humans was endemic in many inland parts of North America – namely the Great Lakes, Appalachians and the Northwest – the so-called 'goiter belt'. Iodine is an essential trace element used in the production of thyroid hormones and in 1924, an iodised salt was made available that led to significant improvements in thyroid health.[90] It is entirely possible that medicinal whisky from Islay, with the smell and taste of iodine, would have been an attractive treatment prescribed by US doctors. Found naturally in elevated concentrations in seawater, in coastal areas and offshore islands, iodine finds its way into local peat, barley and freshwater. For this reason, it is also present in small and variable quantities in whisky, particularly those produced in places like Islay. Scientists have now found that the concentrations in whisky are nutritionally insignificant and it is not possible to taste iodine in whisky at such low concentrations.[91] What is being tasted and smelt are bromophenols and phenols.

Bromophenol is a phenolic compound produced naturally by marine organisms including seaweeds, sponges, shrimps and fish.[92] It has a single phenol ring structure, with two or three additional bromine atoms attached to the outside. Characterised by an extremely low taste threshold, it plays a big part in the flavours of seafood; isolated chemically, it is described as medicinal or iodine-like. In whisky, the highest concentrations have been measured in peated Islay whiskies, notably Lagavulin and Laphroaig, where it is either formed by the natural bromination of phenols or may simply arrive from the ocean in droplets on a sea breeze.

So, what is the truth behind the prohibition Laphroaig story? Officially, of course, no Scotch was shipped to the US during Prohibition. Intermediaries did the job for the big producers like DCL.[68] 'Runners' provided the link between Scotland and the US, using the Bahamas, Cuba, Mexico and Quebec as dropping-off points. It was a form of controlled bootlegging. We know from court testimonies about intercepted whisky that Islay whisky was certainly finding its way into America. In 1925, the *Boston Daily Advertiser* ran a story about a captured 400-gallon consignment of '*Ilay*' whisky that had recently arrived from the west coast of Scotland.[93] In court, the assistant analyst from the Department of Health testified that 'he had never heard of *Ilay*, but that analysis showed this to be the best Scotch whisky out of 2,000 brands which he had sampled'. What sampling method he used was not recorded. Ardbeg whisky was also being shipped to Canada during Prohibition and one can safely assume it found its way across the border to America. But where is the physical evidence? I could find no bottles of prohibition Laphroaig, advertisements, photographs, auction lots, not even a doctor's prescription note for Laphroaig amongst the mountains of prohibition memorabilia. Nothing.

Nobody seems to know for sure if the story is true or not and even those well-respected whisky historians Moss and Hume[94] only go as far as saying that Laphroaig 'probably exploited the America market for medicinal whisky during Prohibition', but they proffer no evidence. I contacted the distillery – nothing. Finally, I resorted to asking the American whiskey historian and expert on Prohibition whiskey Josh Feldman. He replied, 'In my studies of American whiskey history, I've not come across medicinal bottlings of Scotch.'

Marcel van Gils, co-author of *The Legend of Laphroaig* was equally sceptical, 'I have never found any evidence, documents, photos, for the Prohibition story. And I don't believe it is true, just early marketing, probably by Ian Hunter.'

PEAT AND WHISKY

It is a wonderful and an entirely plausible story, and one that is very attractive to some, particularly those who market Laphroaig. But is it fake news? I for one don't always like to see 'the truth get in the way of a good story', but it would be nice to see some evidence of Laphroaig's famous green one-pint bottles and a label with the words 'For Medicinal Use Only'.

Five

'Every Peat has its Own Smoke'

*'Malt whisky is not simply a wonderful drink, it is an expression
of the nature of the land in which it is distilled.'*
Philip Morrice, Schweppes Guide to Scotch, 1983[1]

The End of Peat Terroir in Whisky?

In February 1968, several Scottish newspapers carried the story
that DCL had announced that it was to cease floor maltings at
thirty working distilleries and make over 200 workers redundant.
The list is long, makes for difficult reading and includes names like
Mortlach, Cardow, Royal Brackla, Oban, Clynelish, Knockdhu,
Dallas Dhu, Glenlochy and Glen Garioch.[2] Many were using
locally cut peat in their malt kilns and in the coming years these
bogs, with long established links to their neighbouring distillery,
would fall into disuse. Outwardly, the UK economy was growing
after the post-war recession and the whisky industry was booming
as distilleries restarted, new ones opened, and existing distilleries
expanded. The capacity of a distillery to malt its own barley had
become the rate-limiting step, effectively controlling the volume
of spirit it produced. If a distillery wanted to make more whisky,
it needed more malt, new stills, mashtuns, washbacks and warehouses. It needed more room, but many were short of space,
crammed onto level ground along riverbanks or squeezed into
back streets. Expanding the footprint of the distillery was not an
option. The solution lay elsewhere – the spacious and labour-intensive floor maltings had to go.

In their place arrived a new breed of large-scale processing
plants to supply the growing needs of the distillers for malted

barley, both peated and unpeated. Throughout the 1960s and 70s, large industrial-scale maltings were opened – Burghead and Glen Ord in the Highlands, Glenesk and Pencaitland in the Lowlands. For a while, most of Islay's distillers retained their floor maltings, but in the early 1970s, Lagavulin, Caol Ila and Port Ellen distilleries all stopped malting their own barley, and in their place Port Ellen Maltings was built with its three large peat-fired kilns. In 1979, Crisp Malts opened the Portgordon plant at Buckie in the heart of the fertile, grain-producing region of north-east Scotland.

In the wake of the Arab-Israeli War and a global oil crisis, the UK's economy faltered badly in the late 1970s – boom turned to bust and Scotch whisky producers were left with a growing 'whisky loch', overflowing after years of overproduction. As a consequence, large numbers of distilleries including Port Ellen, Brora and St Magdalene stopped working and were mothballed. Many never reopened and were eventually demolished. But new maltings continued to be built and in 1982, amongst the barley fields of Moray, DCL opened the huge and impressive Roseisle Maltings, later adding a modern distillery complex in 2009. In the first years of the twenty-first century, there are now twenty-six medium- to large-scale industrial maltings operating in the UK supplying different types of malt to the brewing and distilling industry both at home and abroad.[3] Twelve are in Scotland with a cluster in the north-east, and all are in touching distance of major centres of barley production. The one exception is Port Ellen Maltings that relies on grain ships like the *Victress* that week-in, week-out circumnavigate the coastline of the British Isles, bringing a never-ending supply of barley from the granaries of eastern England and the north coast of Europe to be malted and infused with the peat smoke of Islay. Seven UK maltings produce peated barley for the whisky industry (around 100,000 tonnes a year, or 10 per cent of the malt used by the whisky industry) and some, like the

Portgordon Maltings, specialise in producing a super-peated malt using a custom-made peat burner.

The Scotch whisky industry was changing rapidly in the 1960s, '70s and '80s, and floor maltings soon became a thing of the past or, some would argue, a rare, present-day curiosity. The peat sheds disappeared, some finding a new use as a boiler house, tractor shed, or as a home for hardy football supporters in the Highland League. In their place at the maltings of Port Ellen and Glen Ord rose large cavernous modern metal structures that resembled small aircraft hangers, capable of storing enough fuel to feed the needs of the peat kilns. As malt production became centralised, so did peat production. It became focused on fewer and fewer bogs that needed to be both close to the maltings and large and deep enough to produce a reliable annual peat harvest. With one or two notable exceptions, the bond between a distillery and its local peat bog had been broken. The use of peat in whisky making had become homogenised – its place of origin of little importance in a modern age.

This brings us to the subject of terroir in whisky and in the context of this story to peat terroir – did it ever exist and if so, does it still exist? Within the whisky-drinking world, Islay is the place most closely linked to the concept of peat terroir. Its island peats are famously *Sphagnu*m-rich, maritime and even seaweed-rich; its rivers and streams the colour of liquid peat; its smoky whiskies heavy and medicinal, with a whiff of iodine and a pinch of salt. Peat terroir works particularly well for the Islay malts, where a glass of Caol Ila or Ardbeg will spirit you back to that windy, wet, peat-covered island on the west coast of Scotland. It is of course an image eagerly exploited by those marketing the island's whisky. But look closer at the Islay single malt whiskies and there are clear differences. Diageo's Brand Ambassador Colin Dunn summarised them rather well:[4] Caol Ila 'minerally peat, effervescent, grassy, with smoky bacon crisps on the nose'; Ardbeg 'sweet peat'; Lagavulin 'oily peat' and Laphroaig 'dirty peat'.

Driving these differences is both the whisky-making process and, some would argue, peat source. Peat bogs cover around 20 per cent of the land surface of Islay, many found in low-lying coastal areas, others occur inland up to altitudes of 250m. This tells us that they formed at different times and at different rates. Currently, large-scale peat harvesting is restricted to three mosses; Gartbreck, Machrie Moss and Castlehill Moss, while older distillery bogs like Kintour Moss, Duich Moss, Conisby and a large area around Loch Gorm, where Bowmore once cut their peat, have now been abandoned.

Although at one time its whisky was no different to the other Islay distilleries, Bruichladdich Distillery is deliberately different from its neighbours. In the 1960s, the distillery lost its Islay style and switched to making an unpeated malt, attractive to the blenders and the important American market.[5] In the 1970s, 80s and 90s, Bruichladdich was said by Jim McEwan and others to be 'the islanders' favourite dram'.[6] The distillery now produces the famous Octomore, the 'peatiest whisky in the world', but oddly doesn't use Islay peat or malt from Port Ellen Maltings because they are unable to peat at the levels required by the distillery. Instead, they have worked with Bairds Malt in Inverness to optimise a five-day cold-smoking process to produce the super-peated malt. The key to the process is recirculating the peat reek over the malted barley and keeping it moist by spraying the barley with water to maximise its stickiness to the phenols.[7] The peat used to make Octomore once came from the Flow Country in Caithness.

Whisky terroir has become a topical, controversial and divisive subject that polarises parts of the industry, and to many the arguments have become an unwanted distraction. In the blue corner, we have a number of new craft whisky makers seeking individuality, unique flavours and provenance; in the red corner we have the establishment, making and marketing huge volumes of spirits driven in most cases by the needs of the blenders for consistency. A research paper published in 2021 linking local soil, climate and malted barley to flavour compounds found in new-make spirit was hailed by those who funded the study as an end to the debate.[8] Science had provided the proof, and whisky terroir had won the day. Traditionally used by wine makers as a sense of local environment and place, Dr Nick Morgan has called the concept of whisky terroir 'Cultural appropriation of the worst kind'.[9] While it might appear to many as a marketing construct propagated by a new breed of craft distillers, that is palpably untrue – the whisky industry, old and new, has always been very good at promoting its whisky using local climate and the special qualities of its water, barley and peat.

Despite the pursuit of liquid volume and consistency under-pinned by science and process, place has always had a home in whisky making and marketing. This is whisky terroir in the broad-est sense of the term, widely exploited by the those who promote whisky, whatever the size of the distillery. At the root of this divi-sive issue is the absence of an agreed definition of the word 'terroir', a word that should not be confused with provenance, a term that refers to authenticity. If a specific wine, cheese, sausage or smoked ham can be clearly traced to its place of origin, it has provenance.

In his book *The Terroir of Whiskey*,[10] Rob Arnold somewhat understates the issue, calling terroir a 'contested word', with a defi-nition that has evolved over time. He points out that contrary to popular belief, the word terroir 'stems from the Latin word *terri-torium*, which can be roughly translated as territory or an area of land with defined boundaries'. In an attempt to resolve the issue, a legal definition was established in 2012 by the European Union that included both local 'natural and human factors'. This meant that wine terroir, for example, could be influenced by a wide range of factors – soil, climate, topography, geology, ecosystem, local craftsmen, viticulture and grape variety. In short, a wine with distinct terroir should taste of somewhere, not like something.

Rather than drawing a line under the issue, the intervention of the European Union appears to have resolved little. In the whisky world, terroir clearly remains a contested word, some would say a bastardised word; arguably a word that is confusing and best avoided. Does a distillery that has a unique production and maturation process, passed down from generation to generation over the last 200 years, not have local terroir, even though it uses barley imported from Poland or Estonia? Are the distinctive tastes of single malt whisky made at Lagavulin and Caol Ila distilleries on Islay, using exactly the same specification of peated malt sourced from the same Islay bog, good examples of why process defines whisky character, rather than terroir?

The concept of whisky terroir becomes even more intractable when the whisky is aged and matured in wood, where the flavours of the cask become more important over time. For that reason, whisky terroir, if it does exist, should be easier to detect in young rather than old whisky as well as new-make spirit. In recent years, we have seen a birth of terroir-focused distilleries. Westland in the US city of Seattle, Bruichladdich and Kilchoman on Islay, Daftmill and Kingsbarns in the barley fields of Fife. Some speak the language of terroir while others avoid it and just get on with making whisky. On the south coast of Ireland in a former Guinness brewery, the whisky makers at Waterford Distillery have fully embraced whisky terroir with a clear emphasis on barley, naming the farm where it was grown, the soil type and whether it is cultivated using organic or inorganic methods. They have become the standard bearers for whisky terroir and are exploring biodynamics, a farming approach that applies the principles of Rudolf Steiner – wholly organic, locally sourced, reuse of materials – with planting and harvesting times influenced by astronomical cycles. The nineteenth-century crofters of Skye were early pioneers of biodynamics, not only making butter during an auspicious lunar phase but preferring to cut their peats when the moon was on the wane.[11]

Although the word terroir has only recently found its way into the lexicon of whisky, it is not a new concept. Without using the term, almost 100 years ago Stuart Henderson Hastie recognised the regional importance of peat:

Moss and peat ground flora differ from the north and south of Scotland and differences in these plants and their products appear to provide a more logical explanation in whisky type.[12]

In 1930, Aeneas MacDonald drew an early parallel with the wine regions of Europe and mused about the importance of geography in the character of whisky, specifically air, water, peat and barley.[13]

That brings us specifically to peat terroir, if there is such a thing. In the words of the first line of a well-known proverb from South Uist, 'every peat has its own smoke';[14] for peat terroir to exist, the smoke produced from burning locally cut peat should have an influence on the character and taste of locally produced whisky. After demonstrating that peat harvested from Scotland's Highland and Island mosses were chemically different with distinct marker phenolic compounds, Barry Harrison's research went several steps further.[15] Under controlled laboratory conditions, he burnt his different peats, mimicking the pyrolysis that took place in a peat kiln to create a smoky malt. He then replicated the whisky production process of mashing, fermentation and distillation to produce a peated new-make spirit. He was trying to follow the chemical fingerprint of the individual source peats all the way to the spirit. The final stage of his PhD research was a chemical and sensory analysis of the new-make spirit using a panel of trained experts. Barry found that the peated malt retained many of the distinct chemical characteristics of the individual source peats after pyrolysis. In addition, although many of the peat-derived compounds did not survive the production process, both his instruments and the sensory panel could detect the different marker phenolics and flavours associated with specific peats. In short, peat sourced from Islay, Orkney and the Scottish Highlands all had an impact on flavour, or to put it another way, local peat terroir could 'survive' the production process and be detected in the new-make spirit. But for how long do these peaty chemical and sensory markers persist during maturation and can they be detected in single malt whisky aged for five, ten or even twenty-five years? It is an open question, although many of the specific peat-derived congeners have low sensory thresholds compared to other flavour compounds, allowing them to be sensed at very low concentrations that might stand the test of time.

Scotland and its five officially designated whisky regions are all to a certain extent loosely linked to inherent differences in peat character, created by regional variation in climate and flora. Islay, Campeltown and Orkney are all places with maritime peat occurring mostly in low-lying coastal areas. The Highlands and Speyside are historically linked to non-maritime, inland peat deposits, many forming at altitudes over 300m above sea level. Further south in the bogs above Falkirk and Slamannan that at one time supplied the Lowland distilleries of Rosebank, St Magdalene and Bankier, the peat is more mineral-rich. Akin to barley growing on different soils, the fact that thousands of years of peat growth in geographically distinctive parts of Scotland produces a chemically different material is not a surprise.

Unearthing evidence of peat terroir and the relationship between a distillery and its local bog takes you on a journey into the past. Guided by books, newspaper articles, distillery records and archives, word of mouth, a photograph on the wall of a visitor centre, or even a few words on a bottle, tracing the past links between a distillery and its peat can be a tricky path to follow that sometimes requires feet on the ground. In the archives of the University of Glasgow is one of the most remarkable records I came across in researching this book. Four pocket-sized *Peat Cutting Books* covering the period 1885 to 1963 list page-by-page, person-by-person, peat bank-by-peat bank, the winning of the peats at Bunnahabhain Distillery on the north-east coast of Islay.[16] It starts four years after the distillery was built and ends on the year that the owners stopped floor malting their barley. It is not only a vivid record of every peat cut during the lifetime of the distillery, but it also charts a fundamental change in the character of Bunnahabhain single malt from a robustly peated to a largely unpeated spirit. The unit of peat measurement was the 'perch' and there was an expectation that individuals cut around fifteen to twenty five-metre-square perches a day. A carefully drawn map of *Bogach nan Tarbh*, the Bog of the Bulls, accompanies one of the books, showing thirty individually numbered peat banks laid out in parallel groups, the peat-cutters' road in red ink along with their hut. The individual banks are still evident from a satellite image that overlays perfectly onto this hand-drawn map.

In its early days, Bunnahabhain was one of the largest and most profitable distilleries in Scotland and it needed huge amounts of peat. The first names in the book, Neil M. McPhee and Allan M. McDonald, were paid a total of £39 16s 3d for cutting 1,741 perches in the 1885 season, the highest number ever recorded. They received 9d per perch to cut and 1d for laying. In the following years, many names, beautifully scribed in black ink, appear and disappear from the pages; Archibald Gibb and 'boy' for cutting, Mrs Campbell for 'cleaning banks', schoolboys and 'own men' for 'peat footing and refooting', and 'one man and a horse'. Other tasks, all paid at different rates, included making and repairing the peat moss road and bridges, cleaning ditches, carrying peat in sacks and carting. Women were a significant part of the workforce and in 1924, twelve women were paid a total of £11 11s 0d for cleaning the peat banks. By then, the distillery had started to make an employer contribution to their workers' insurance. And in the final pages of the fourth book, a record of the last person to win peats for the Bunnahabhain Distillery: John McIntyre & Co. cut 888 perches in the 1963 summer season.

While on the surface the numbers recorded in the *Peat Cutting Books* are a simple inventory of the annual peat harvest, underlying the figures are significant periods of socio-economic change. The numbers are rarely static, declining during wet summers and recovering during drier years. When men were called away to war, less peat was cut, but scarcity of coal during two post-war periods created a surge in demand for peat. Few peats were cut during periods of a significant downturn in production and years of dormancy are recorded as blank pages.

In the mid-1880s when Alfred Barnard toured the distilleries of Britain and Ireland, he visited a total of more than 120 whisky distilleries in Scotland.[17] His writings may be incomplete and need a bit of research to fill in the blanks, but at least seventy-five of the rural distilleries he visited were using local peat sourced from a minimum of sixty-six individually named bogs. The big city distilleries of Glasgow, Edinburgh and Aberdeen as well as Campbeltown were able to use their ports to ship in peat from afar, with the islands of Eday, Lewis and the Shetlands being important suppliers. But these were exceptions and demand for local peat continued to grow further in the 1890s with the building of thirty new distilleries in the Highlands, many in Speyside. New bogs were opened up and at the turn of the century, around 100 distilleries were using local peat to dry their floor-malted barley. Some of the distillery-bog links are still celebrated by a name on a bottle – Glenlivet Faemussach, Benriach Birnie Moss and Highland Park Hobbister. Whisky terroir, its flavours influenced by local peat, local water, local barley and local craftsmanship and culture, was probably never stronger.

Whilst proximity to a local peat moss was advantageous to many distilleries, others were less concerned about the origins of their peat, as long as the supply was secure and they paid the right price. One of those was Glenmorangie, a giant of the modern whisky world synonymous with a light, fruity, sweet, cask-driven style of single malt whisky. But it was not always like that. When Barnard dropped in, it was just about to be pulled down and rebuilt.[17] He spent little more than half a page chronicling this famous distillery, concluding it was 'the most ancient and primitive we have seen'. He wrote, 'Peat was the only fuel used in the establishment', brought in by rail from Sutherland, later by ship from Eday and locally from peat contractors in Edderton.[18] Glenmorangie bought its peat on the open market, driven by price, quality and security of supply. In 1887, the rebuilt distillery was using steam-heated stills, the first of their kind in Scotland. After the end of World War II, coal trains provided the main source of fuel for the distillery's boilers, but until the maltings closed in 1980, peat continued to arrive by road from Aberdeenshire. By then, the coal-fired boilers had been switched to heavy fuel oil, the power requirements of the distillery doubling as the number of stills increased from two to four. In the future, the distillery's twelve stills will be run on natural gas, including biogas produced from its own anaerobic digestion plant. There is no better example of how a distillery has adapted and evolved in response to successive energy revolutions and markets.

Since the turn of the twentieth century, the number of active, working bogs supplying the big industrial maltings and the few remaining distilleries that still fire their own peat kiln has shrunk dramatically. At the last count, there were just seven or eight working bogs in Scotland supplying the needs of the distillers. Four of the bogs are on the peat-rich island of Islay, one each on Orkney and Lewis. The most important, St Fergus Moss, sitting on the low-lying coastal flatlands close to the village of New Pitsligo, has become the current epicentre for the production of peat destined for the whisky industry. What follows is the remarkable story of a family-run business lying at the very heart of smoky whisky that at the beginning of the twenty-first century supplies peat to distillers and maltsters in Scotland and around the world. If you are drinking a blended or single malt whisky with a splash of peat or big smoky flavours as you read this, it could be from Islay or Orkney, but it is more likely that the provenance of the peat is Aberdeenshire.

'Aberdeenshire has got a new industry.'

In 1907, the local press announced the registration of a new business, the Northern Peat & Moss Litter Company Limited, based in the village of New Pitsligo.[19] It was a peat harvesting operation that would use methods never previously seen in this part of Scotland, where a predominantly rural farming community cut its peats by hand and barrow using age-old, traditional practices. Removed from the demands of producing peat for large towns or major centres of distilling, Aberdeenshire's extensive peat mosses were largely untapped. When the company was first registered, several distilleries on Speyside had begun to burn peat from the mosses around New Pitsligo. In the mid-1880s when Alfred Barnard visited the coastal distillery of Glenugie in nearby Peterhead, he commented, 'The peats used are of excellent quality, and are to be found in great abundance about five miles distant from the works'. These were the formative years of the link between New Pitsligo, its surrounding peat mosses and the whisky industry that with time would become stronger, and by the beginning of the twenty-first century would make this area the most important player and supplier of peat in the whisky world.

Main Street New Pitsligo is wide and stretches out like some small town in the American Midwest. There is no tumbleweed, only rows of austere grey granite buildings strung out and set back from the road. There is something permanent and unmovable about this Aberdeenshire village – it is easy to imagine it being here when humans no longer walk the earth. The harsh lines of the formidable buildings are adorned and softened by hanging baskets and well-tended flowers that sway silently in the afternoon breeze. For Sale signs adorn many of the properties, but there appear to be no takers. Today, an eerie quietness hangs over the place, with the distinct sweet, aromatic smell of farmyard manure hanging on the breeze. Breaking the silence, a passing tractor and empty trailer come crashing through town, rocking the colourful baskets until calm is once again restored. It is late August, harvest time, and farmers are at work in nearby fields. Another vehicle, heading north this time, a large white van with small windows carrying returning inmates to Her Majesty's prison at nearby Peterhead. The village's chippy is going strong, as is the convenience store and the bar at the Commercial Hotel. Number 40 High Street, once the office of the Northern Peat & Moss Company, is now home to the Sunshine Chinese Takeaway, a seemingly temporary, modern, ugly prefab. A large church, village hall and a beautifully kept War Memorial bear testament to an important past, sacrifice and remembrance.

New Pitsligo is a village built from 450 million-year-old Aberdeenshire granite, and its High Street reflects its former strength and wealth. The village was paid for by granite, and at one time there were five granite quarries noisily blasting and dressing stone within one mile of where I am standing. Its well-made granite buildings, with their chiselled ornate flourishes and facings, were built to stand the test of time. It was founded in 1787 by Sir William Forbes and constructed to a designed plan with a laid-out grid of streets on either side of the High Street. It is named after Lord Pitsligo, the original landowner, who, like so many in these parts, backed the losing side in the 1745 Jacobite Rebellion and lost his estate. The village was at one time famous for its lace as well as its granite, but all around are peat bogs, and as work in lace and quarrying granite declined, peat harvesting became an important source of work and income for the people of this village.

Up until the early 1900s, peat had always been the principal fuel for the small communities of north-east Scotland and each year, a small fee was collected by the landowner from each family for the right to cut local peat from the home bank to heat their farmhouses and dwellings. The tradition lives on to this day and just outside New Pitsligo at Lambhill Moss on the long straight Strichen Road, the old ways are still kept alive by the few. That August morning, it was bright and breezy out on the bog and the mossers had recently been at work – their boot prints still fresh in the wet exposed peat. They had stripped back and discarded the upper fibrous, mossy layer and then cut down twice into the dark, rich peat below. The freshly cast peats were neatly stacked in small ricks of ten turfs each, drying in the warm breeze blowing across the moss that today was strong enough to keep the airborne bugs at bay. Branches of recently exhumed silver birch lay scattered all around, the bark iridescent after thousands of years below ground. I scoop up a handful of rich peaty water from the dark pools that form at the foot of the cut peat bank; it is soft, smooth and tasteless. Peat-filled bags lie littered around and makeshift paths of wooden boards criss-cross the wet floor of the bog and bridge water-filled drains, allowing the mossers to wheel their barrows of cut and dried peat to the Strichen Road. Some of the original deep drainage ditches are still active, others in a state of disrepair. I come across a well-made peat barrow, constructed from Perspex and metal, with a single rubber front tyre and two handles; a modern take on an ancient design, a window into the past. Three mossers pay £40 each year for the right to cut 100 barrowloads of peat on the Strichen Road. The rate has remained unchanged for the last twenty-five years.

Aberdeenshire's newest industry got off to a slow start in the early years of the twentieth century. Eleven local men originally employed by the bosses were not up to the job and were sacked, or as the local newspaper put it, 'requested to make room for others who would accomplish the amount of work more in accordance with the manager's ideas. Their successors are four Dutchmen.'[20] By 1914 and with the early problems resolved, peat extraction from the mosses of New Pitsligo had become a large and successful business enterprise and its peats had already become much prized by distilleries across the world. In 1909, a steamer left Aberdeen with 1,000 bags of New Pitsligo peat onboard, *en route* to a distillery in Australia via the Port of London.[21] In 1912 and 1914, the shipments to Australia were repeated; fifty tonnes of 'best class' New Pitsligo peat was sent to the whisky makers down under.[22,23]

Radiating out across the open bogs were now miles and miles of light railway tracks constructed to bring in the peat, as the traditional labour-intensive old ways of barrowing and wheeling had been abandoned and replaced by the Dutch Method. This involved casting a two-foot-square of peat, called a 'gullet', from a straight peat face 200-yards-long, hand cutting it into blocks and leaving it to dry on top of the bank. The half-dry peats were then stacked before being loaded onto bogies or trucks, often by women, and then transported by light railway to the works for processing. Brushwood was laid between the rails to allow a large cob pony to haul the three-ton truckloads of peat without sinking. Dozens of long, straight ditches were cut through the bogs, each with its own piece of parallel light railway, that were widened and deepened as the peat miners dug down into the moss. Lower down they found treasure...

> ...the fourth layer or bottom peat is the most valuable of all. It is what is known as the distillers' peat and is extensively used in the manufacture of whisky. A representative of the Pitsligo factory recently went on tour through the northern distilleries, and from Old Meldrum [home of Glen Garioch Distillery] to as far north as Inverness, and sold no less than 800 tons.[23]

When one moss was mined out, the peat workers moved on to the next one, taking their tools and railway tracks with them. By 1914, the Dutch workmen and manager had moved on and the local mossers had given up on the old ways and learnt and adopted the more efficient and faster new ways. I had heard the very same story elsewhere in the St Laurence Bar in Slamannan about peat harvesting at nearby Fannyside Moss – 'It was the Dutch that started it.'

As war broke out in 1914 and Europe descended into chaos, local industries and resources took on an even greater importance. The peat industry in this rural part of Aberdeenshire was booming and the village of New Pitsligo became a centre for the harvesting and processing of peat. The Northern Peat & Moss Company was prospering – it had a board of directors and supportive shareholders, and each year they set out for the mosses for a good lunch, followed by the annual inspection with the day rounded off by a 'capital picnic-tea'.

> The industry provides work all the year round for many hands from the adjoining village of New Pitsligo. The annual inspection of the Pitsligo Moss Works and Factory in 1917 started

with lunch at the Pitsligo Arms Hotel. There, the directors and shareholders were joined by Mr D. M. Godsman, managing director of the company and Mrs Godsman.[24]

David Mackie Godsman was not one of the original directors of the company, but his family bought into the business after the original owners apparently went bust. The well-kept Pitsligo Arms Hotel where Mr Godsman and his wife Jeannie met their guests for lunch still stands, but when I visited it was closed again after a brief revival. I had hoped to stay the night, but all was quiet and there was no sign of life or imminent arrival. After their lunch at the hotel more than 100 years ago, Mr Godsman and the inspection party headed out to the moss works and soon arrived at their destination:

The factory is a three-storeyed, corrugated iron building, on the ground floor of which are the engines which drive the machinery and the arms and levers in connection with the pressing of the bales of moss litter ... all the railways throughout the moor converge to the factory. As the bogies come in laden with peat into an elevator it is sent right to the top storey of the factory, where it is crushed, or rather torn into shreds.

The tearer was known as the 'devil'. Adapted from the textile industry, it consisted of two large cylinders fitted with sharp teeth that revolved in opposite directions. On a good day, the devil and the mechanical baler were capable of processing fifty tonnes of moss litter.

The day is now warm and that afternoon I approached the now silent moss works down an overgrown, rutted track. Open water covers some of the abandoned peat workings, now home to a gang of metal-blue dragonflies, which stop briefly to examine me before darting off in search of more manageable prey.

Deep drains still transect much of Middlemuir Moss, still actively moving water away from this squashed and sunken landscape, a fitting testament to the skill of the Dutch engineers, but a barrier to *Sphagnum* growth and future peat renewal. All around, the original peat extraction cells are clearly picked out by the lines of purple heather growing in the driest areas raised high above the current water table. Jutting out from the peatland are huge rounded granite boulders, dislodged by the peat workers, remnants of a past glaciation and evidence that deep below my feet, beneath soft peat and wet clay, lies solid grey Aberdeenshire granite, the rock that built New Pitsligo.

The old moss works is boarded up, tidy and intact. Through a small hole in big rolling doors, I spy a loop of railway tracks set into a concrete floor where the peat wagons once ran. Round the back is a door to an upper level where high-quality fuel peat was stored. The peat wagons rolled off the moss along three elevated fingers, converging and arriving at the works through a small cutting before disgorging their loads from high on a ramp. Like a child's model railway on a carpet of moss, the tracks were movable and could be broken up and reassembled, extended and shifted when an area became exhausted and a new one opened up. The peat wagons were moved by manpower, then horsepower and finally by diesel power. In 1978, the Northern Peat & Moss Company's peat sod railway ceased operation and the last of the diesel locomotives were sold off. The factory finally closed in the late 1990s, but Ian Birnie, a local who has farmed here for thirty years, told me he can still remember the peat factory in operation and the lorries and wagons lined up in the morning to take away the processed peat. As I talk to Ian, perched high up on his new tractor, he tells me he still hauls lorry loads of dry peat for Mr Godsman. As he smokes one cigarette after the other, I ask him if he burns peat at home. 'No, I hate peat smoke – it makes my eyes run.'

The moss works was one of two peat-processing factories in the area and as the use of moss litter for animal bedding declined, the company increasingly focussed on selling the shredded peat to the horticultural market. The second works was at Lambhill and produced mostly fuel peat. Now it is an animal sanctuary and I had hoped to visit, but the gate was drawn across the track and a sign stated 'Closed today, due to staff sickness'. The Godsman family business was one of several large peat companies that operated in the New Pitsligo area throughout the twentieth century, in what was a golden age for peat production in this part of north-east Scotland. Their competitors, the Middlemuir Peat and Moss Litter Company and the Irish-German owned company Herbst, are long gone, and only the Northern Peat & Moss Company remains. Its main customer is now the whisky industry, but in the late 1980s and early 1990s, a new and unexpected export market opened up for Scottish peat on the other side of the North Sea.

Sweden had decided to phase out its nuclear power stations and move towards a future built partly 'upon indigenous sources of electricity generation, notably peat and wood chip'. For several reasons, the Swedes decided not just to use their indigenous, home-grown peat but looked overseas, and in early 1987, the first cargo of Scottish peats was shipped out of Wick by the Highland Peat Company.[25] They were cut and harvested from Dale Moss in Caithness, 25km (16 miles) inland from Wick Harbour. A three-year contract was signed with Highland Peat and its sister firm, the Northern Peat & Moss Company of New Pitsligo. Mr Miller, the owner of the company, said the shipment of 1,580 tonnes of peat to Gothenburg was 'the biggest export cargo, which has gone from the port. This is the first time Scottish peat has been exported in bulk and we are obviously hoping that we will be able to do regular business with the Swedes once our present contract is completed.'

For a short while, the east-coast ports of Wick and Peterhead became hubs for the export of peat, and heading across the North Sea were the largest shipments of peat ever to leave the shores of the British Isles. In one year, 30,000 tonnes of peat were sent to Sweden with a capacity to generate 100,000 kilowatt hours of energy. Peat from Caithness and New Pitsligo was used to power small local boiler plants in Sweden and generate hot water. Sweden is a peat-rich country and at the time burned significant amounts for domestic power generation, but they needed more, and the tightening of environmental regulations at home made Scottish peat commercially attractive. Some of it was mixed with domestic waste to generate heat for local hospitals, houses and schools. But the excursion into offsetting a fuel shortage in Sweden caused by a decision to move to a non-nuclear future by using alternative energy sources, including Scottish peat, was over almost as soon as it began. Government policy and priorities changed and the trade in peat from Scotland to Sweden came to a shuddering halt. In 1990, the Highland Peat Company was put up for sale and Caithness and New Pitsligo peat no longer crossed the North Sea. A small peat operation continued for a while in Caithness, but the focus of peat production was moving progressively south towards New Pitsligo.

'Ye canna beat a peat heat!'

I had arranged to meet Lesley Craig at the company office at 9.30am. Driving down the east coast road from Fraserburgh, home to Scotland's largest deep-sea fishing fleet, I pass close to the old Cold War listening site on Mormond Hill. Once run by the US military, its rusting masks and satellites dishes looked out across the flat coastal plain and scanned the airspace above the North Sea for incoming Soviet aircraft and missiles. This morning, the fields are busy; Aberdeenshire is in a frenzy of late-August harvesting as farmers go all out to get the barley combined and safely stored away. The weather is on the change and tomorrow the forecast is for heavy rain.

Agriculture and energy dominate the landscape of this part of north-east Scotland. Emerging from behind the coastal dunes are the steel towers and yellow flares of the St Fergus gas terminal, the point where a quarter of the UK's gas comes ashore from the natural gas fields of the North Sea to be processed, before being sent back underground in pipelines towards the industrial heartland of Scotland. Before reaching Peterhead, I turn inland, away from the coast, and drive towards three large wind turbines rotating steadily in the wind, their concrete bases sunk deep into black earth. As I get closer, I stop and get out of the car. Across the harvested fields of barley are bare fields of peat, an expanse of blackness with small conical hillocks and long drawn-out ridges of harvested peat stretching out across a slight rise in the land. Lying 5km (3 miles) inland from the coast and just 40m above sea level, this is St Fergus Moss and I am heading to Blackhills Farm, the current home of the Northern Peat & Moss Company.

The company is owned by Neil Godsman, the third generation of a farming family who have been harvesting and selling peat from the bogs of New Pitsligo for more than 100 years. Today, like other local farmers, Neil is away bringing in the grain harvest from one of his farms, but there is little that Lesley Craig, who has managed the company's office for the last twenty-four years, doesn't know about the Northern Peat & Moss Company. After a thorough welcoming examination by Jimmy, the company's Border terrier, that felt more like a full body search at airport security, I was given the all clear and we set off for a tour in Lesley's white SUV. This year's harvest had already been completed and out on the moss all is silent except for the three white turbines relentlessly churning away in the wind. The peat is extracted from the bog during the month of April and then dried on the surface for another four weeks before being taken off the moss. Compared to the old ways and the Dutch method, this is a revolution in terms of efficiency and time, and for most of the year the company's tractors, ditchers and harvesters lie idle.

The Northern Peat & Moss Company owns 1,000 acres of prime Aberdeenshire peat bog, actively working a small part at any one time. Over the years, the Godsman family has harvested peat from Red House Moss, Craigculter Moss, Lambhill Moss and now St Fergus Moss, moving from bog to bog extracting peat from the area for heating and distillery malting. The company's motto is 'Ye Canna Beat A Peat Heat!' Peat has been cut from the moss where I now stand since 1977, but it was only after Neil Godsman purchased St Fergus Moss from Herbst in 1997 that the operation was scaled up to its present level. In 2012, the company was given a twenty-year peat extraction licence with agreed environmental guidelines that allowed them to cut down to within half a metre of the base of the moss. How much they extract is governed by how much they can process and sell; typically 7–9cm a year is removed from about half the area of the moss. Unlike the Islay peat mosses, where year-on-year peat production can be affected by the weather, St Fergus Moss is in the rain shadow of the Scottish Highlands in the cool-dry east of the country and so largely escapes the wet years that can halt peat production out west. In 2032 when peat extraction comes to an end at St Fergus Moss, the company will enact an agreed aftercare plan that involves blocking the drains and re-profiling the surface. The site will then rewet, moss plants return and recolonise the bare earth. St Fergus Moss will start to regrow its peat and once again become a site for capture and storage of carbon dioxide from the atmosphere.

Peat started forming at St Fergus in a sizable post-glacial lake at the end of the last Ice Age. Over the years, as the lake was inundated by wetland plants and slowly shrunk in size, it created a 420-acre, dome-shaped acidic bog. The pollen record shows the transition between the Bronze and Iron Age, as man accelerated the clearance of the local forests around 1000BC.[26] Lying below our feet at one time was seven metres of high-quality peat, the result of 10,000 years of progressive accumulation of partially decomposed organic matter.

The first stage in extracting peat from any bog involves installing a network of drains to initiate drying. When the peat is sufficiently workable, one of three methods can then be used to harvest the peat; namely, cutting blocks or turfs by hand or machine, milling or shredding the upper layers, and lastly the method used at St Fergus, surface compression followed by extrusion of the sub-surface peat as sausages or nuggets. Milling is the most common large-scale extraction technique producing a fibrous peat for the horticulture sector, or a fuel for peat-fired power stations. The extraction process involves levelling the surface, milling or shredding the upper 15–20mm of peat using a machine with rotating drums fitted with pins pulled by a tractor, and then repeatedly harrowing and turning the milled peat until it reaches a moisture content of 50 per cent. On average, twelve crops of milled peat are harvested during the year, although this is highly weather dependent. Like a field of wet hay, the loose peat is left to dry, periodically turned, and when it reaches the correct moisture content, swept up into long ridges and collected by a huge vacuum harvester. Milling is a fast and destructive method that has never been used to extract peat for the Scotch whisky industry.

The peat harvested at St Fergus is an annual single crop made possible by deep drainage and north-east Scotland's relatively dry climate. The Dutch method of drainage is still much in evidence and the company uses a wedge-shaped ditcher that cuts and maintains the deep drains that run across the moss exposing the sticky, grey clay that lies beneath the peat at the bottom of the ancient lake. Each year, the drains are cleared and deepened as the harvesters go deeper into the moss to expose and exploit the older layers of peat. Beneath the clay and scattered about the moss at St Fergus are fragments of bedrock, here a grey Dalradian schist. In a nice piece of geological symmetry, they are the same metamorphic rocks that lie beneath Diageo's Castlehill Moss on the west coast of Islay, where peat is harvested for the local maltings at Port Ellen.

Based on more than 100 years of experience passed on from one generation of the Godsman family to the next, the process for winning the peat at St Fergus has evolved into a highly efficient, well organised and mechanised operation. Firstly, the surface of the moss is tilled and scarified, removing vegetation by a machine called a screw leveller pulled behind a multi-wheel tractor. It consists of a horizontal rotating Archimedes screw that scrapes off any vegetation from the surface and throws it to one side, creating a perfectly flat surface that is ready to be harvested. A compressor and extruder, basically a 'large sausage machine', then scoops up moist peat from 6–8cm below the surface into a large tube with a rotating screw. The extruder moulds five rows of smaller, moist sausages and throws them onto the surface. They fall beside the machine like wet pieces of plasticine, breaking into manageable chunks when they hit the ground and begin drying. These are the peat sausages or nuggets that visitors get to handle when they visit Scotland's distilleries.

The sausage machine is the only specialist piece of equipment used by the company – all the others are basically pieces of adapted farm machinery. They are allowed to dry, turned periodically and then gathered up into long 'wind rows' that form low ridges across the bog surface that helps them reach a moisture content of 45–50%. Then they are collected using a machine that looks like an onion harvester that transfers the sausages by conveyor into a high-sided trailer, dumped into large mounds and driven by tractors to the edge of the bog where they are piled up into vast A-shaped stockpiles using a digger. The careful stockpiling ensures the peat can be stored for many years without getting wet as snow and water runs off and wind keeps the surface dry. From the stockpiles, the peat is fed into a hopper where they are graded or screened as and when required by Lesley's customers. This separates the dust or the dross from the sausages, which can be bagged and sold as 'burning nuggets'.

The whole operation speaks to me of a peat farm, harvesting sods from just below the surface like potatoes, using farming knowledge and modified machines pulled by tractors adapted to working on soft ground. On our way back to the office, we stop to inspect an array of parked, abandoned, gutted and modified machinery, including bits of sausage machines, ditchers, cutters and onion harvesters. All were bought locally, apart from the specialist compressor-extruder that was imported from Sweden.

The Northern Peat & Moss Company harvest between 6–7,000 tonnes of dry peat each year from St Fergus Moss, and Lesley estimates that at current extraction rates there are about fifteen years of peat left below ground. Nowadays 3,500–4,200 tonnes, about 60 per cent of the annual harvest, goes to the whisky industry; the remainder is bagged and sold for domestic fuel. By comparison, Port Ellen Maltings in Islay consumes 2–3,000 tonnes of dry peat a year. The highly absorbent dusty dross, a by-product of peat nugget production, is sold on and used by blueberry growers as a soil improver or, in the past, to chicken farmers. When straw prices are high, it is purchased as an alternative form of cattle bedding. Nothing is wasted. Even as the popularity of peated whisky has steadily increased, they have not seen an increase in demand from the whisky industry and the malting companies. Kilning efficiency has improved, as it must do for a material that is becoming increasingly difficult to source. As the older peat mosses became exhausted and operations shut down, the Godsmans' company has become the dominant supplier for the whisky industry, with sales growing relative to all other markets.

In the past, distilleries were often quite particular about the type of peat they burnt in the malt kiln, so back in the office I asked Lesley if there was a specific type of peat preferred by the industry?

Not really. The maltings are our main customers and they differ in the type of peat they require, some like extra dross added to dampen the fire and some just want unscreened peat as it comes off the stockpiles. Others like a wetter peat to generate more smoke, others just want the dry screened nuggets. Some customers change what they ask for each year. There doesn't seem to be such a thing as malting peat as all the maltsters like different blends and use different types of kilns and systems. Peating malt barley seems to be more of an art than a science but once they have worked out what type of peat suits them then we do our best to supply them with a consistent product.

The company has provided New Pitsligo peat for the large maltings since they were first built in the 1960s. Many like Glen Ord, Roseisle, Portgordon, Buckie, Burghead and Bairds are local, but others lie further afield. 'At times when the year has been wet in Islay, Port Ellen Maltings become one of our biggest customers. In north-east Scotland, we have the perfect weather to cut and harvest peat, much drier than on the west coast. We also drain our mosses well, which can't be said of some of the Islay bogs.' The company advised DCL when they were setting up the peat harvesting operation at Castlehill Moss on Islay, in the aftermath of the Battle of Duich Moss. Substituting Castlehill peat with St Fergus peat might appear a game changer, but the difference might not be as great as it seems – both lie just three miles from the sea.

From deep inside a filing cabinet, Lesley starts to unearth old paperwork – price and customer lists from years gone by, written up on an old ribbon typewriter. They show the huge outreach of New Pitsligo peat in the UK; large-scale maltings from Inverness to Edinburgh and as far south as Gloucester and Essex were all customers. In 1976, New Pitsligo peat was sold for £10.50 per tonne, in 1980 £15.95, and when I visited in 2019 the distilleries and maltings were paying £63 for a tonne of screened or £55 for the unscreened peat. In 1840, hand-cut peat from the Orkney island of Eday was fetching just £0.75 a tonne on the dockside at Kirkwall Harbour.

Lesley's papers show that despite the demise of many distillery maltings in the 1960s, in 1976 the company was still selling New Pitsligo peat to Glenmorangie, Balblair, Balvenie, Glendronach, Knockando, Glen Garioch and the distilleries of Rothes. By this time, the DCL distilleries had gone over to centralised maltings, but peat, or peat and coal, was still being used to dry and add flavour to the floor-malted barley. By 1980, Glendronach and Knockando distilleries had dropped off the list. Glen Garioch continued to use New Pitsligo peat to make heavily peated malt right up until 1995 when the distillery closed. The 1995 vintage was apparently the last Glen Garioch to be made with smoked malt barley; two years later, the distillery reopened, but the whisky from then on was unpeated.

Distillery floor maltings are now largely a thing of the past, but the company still sends two articulated lorry loads of St Fergus peat per year to Springbank Distillery in Campbeltown, a 960km (600 mile) journey from one corner of Scotland to the other. 'We also deliver peat to The Balvenie for their annual Peat Week run. They require just four tonnes a year, so are our smallest customer.'

By the mid-1980s, the company started exporting distillery peat to Japan, and the trade continues to this day. New Pitsligo peat was also being shipped to Australia as it had done in the past. There were other new foreign destinations as well: 'We sell to distillers in India and have been doing so for thirty years,' said Lesley. 'One of our customers is Amrut. I once bought a bottle of their peated single malt at some considerable expense.' She found it a bit 'rough'. 'We also sell our peat to Hillrock Distillery in the USA and Castle Malt in Belgium.' Lesley's favourite whisky is Laphroaig.

Where this story of a family-run business that has operated for more than 100 years in both local and global markets will end is uncertain. At one time, dozens of distilleries directly depended on lorry loads or shipments of New Pitsligo peat. The maltsters are now the middlemen, supplying the flavours and smells of Mr Godsman's peat in the form of peated malt barley to an even greater number of distilleries across the world. Methods have changed, customers have changed, rivals have come and gone, environmental regulations are tighter now than they have ever been. The whisky industry rightfully celebrates its longevity, its foresight and is particularly proud of the survival instincts of its family-run distilleries – William Grant & Sons (Glenfiddich, Balvenie), the Springbank Mitchells of Campbeltown, John & George Grant (Glenfarclas) and the Irish Teeling whiskey dynasty. It should also be celebrating this remarkable Aberdeenshire operation that after small beginnings and riding out the economic waves of the twentieth century has now become the foundation on which the global trade in distillery peat and peated malt is built. But for how much longer will the Northern Peat & Moss Company[27] provide whisky drinkers with the flavours of peat they all crave? Will we need to go further afield for our peat in the future?

Svensk Rök

This morning, I received an unusual package in the post from Sweden accompanied by a note: 'Hi, Michael! Here is a piece of the Karinmossen peat that we use.' Inside is a sample bag containing a black, wet, compressed peat sausage. I let it dry for a few days and then it is ready for examination. It has a fine-textured earthy-brown matrix containing 1–2cm pieces of wood and bleached strips of birch bark. With my hand lens, I can see the remains of black needle-like rushes, fine root hairs and numerous twigs of a common bog shrub. I am taken by the organic purity of this piece of peat and have to search hard to find one or two lustrous flecks of angular mineral quartz grains shining in the light of my desk lamp. And then peat treasure: the broken carapace of two beautiful, iridescent bog beetles, dazzling, as their decorated armour reflects my beam of light into complex interference patterns. Made of chitin, their protective shields have survived perfectly preserved for thousands of years in a wet, cold, oxygen-starved Swedish bog. There is more – the shiny brown oval of a 2mm seed that breaks open when I crush it with a needle, revealing an empty cavity inside. I taste the peat – is it slightly salty?

Sweden and Scotland are both peat-rich nations that have a long tradition of curing fish and meat with smoke, but with the exception of the Gothenburg Yeast & Spirit Factory, only the Scots had exploited their vast resources of home-grown peat to make whisky. That all changed in 1999 when a group of friends and engineers built the Mackmyra Distillery in woods near the town of Gävle. The distillery, named after the village in which the original distillery was located, is derived from a compound word that translates from the old Swedish, *mack* or *macken* meaning 'small bug', and *myra* meaning 'mire or bog'. The little bug on the bog distillery. It is noteworthy for all sorts of reasons – casks are stored 50m below ground in what was at one time an iron ore mine that for a short while became a mushroom farm. The warehouseman checks his casks by bicycle. In 2011, the ingenious Swedes built a second distillery – a 35m-tall climate-smart gravity distillery on the site of an ex-military shooting range.

In 2013, Mackmyra released *Svensk Rök*, Swedish Smoke – a single malt whisky made with Swedish peat or *torv*. At the time, no peated malt was being produced in Sweden, so the owners of the new distillery had to make it themselves. The original proto-type Mackmyra smokery was rather quaint and looked like a large brightly painted dog kennel, with a separate maltings box. The scenic lakeside location was later succeeded by a larger, utilitar-ian smokery – a converted shipping container that both dries and smokes the barley from the maltings floor next door. In charge is Håkan Ekström, a former office worker, 'the distillery smoker', who oversees the production of around eighty tonnes of peated malt a year. A wood fire is lit under the peat and this starts a thirty-six-hour-long smoking regime that ends with a final drying of hot air. The aim is to produce a really dense smoke and achieve a high 60ppm phenol level in the grain.[28]

The owners of Mackmyra wanted *Svensk Rök* to have a uniquely Swedish character so they experimented with seasoning the peat by firstly testing the resinous flavours produced by pine needles, before settling on locally collected juniper twigs. Juniper is one of the tastes of Sweden and has been used for generations to flavour food. Small amounts are placed on top of the peat fire during the smoking process. Mackmyra produce all their own peated malt in this way and the final matured whisky has a distinct dry, spicy, aro-matic smokiness with herbal flavours and a subtle mineral saltiness that speaks of local peat terroir. The source of Mackmyra's distinc-tive peat is the Karinmossen or Karin Bog located to the south of the distillery in a low-lying area 30km (19 miles) inland from the Baltic Sea. It is a type of peatland that rarely occurs in Scotland, and that makes its use in whisky making rather exciting. As my sample demonstrated, it has a saltiness, and that is down to its origins.

At the end of the last and final glaciation, the land that would become the Karinmossen was completely submerged as the swollen Baltic Sea invaded large areas of what is now mainland Sweden. As the effects of the Ice Age receded and the weight of ice was removed, isostatic rebound caused the land surface to bounce back. It is still rising, and each year Sweden grows in size, but at a slower rate than in the past. Although the Karinmossen is now 70m above the current sea level, it carries a saline legacy of the past and in many inland depressions new lakes and wetlands were created that provided the perfect conditions for peat to form on hard Swedish granite.

There are other reasons why Swedish and Scottish peat differ. The Swedish climate is both drier and colder with a quarter of all precipitation falling as snow, typically lying for 100 days a year. This means that the peats grow at different rates, are botanically distinctive, with more aromatic herbaceous plants and sparse conifers that grow better in drier climates. This part of Sweden is also famous for its iron ore mines and smelters, and both timber and peat were important fuels for metal working. Much of the peat and forest land was extensively drained 100–200 years ago, drying out the low-lying wetlands and creating the right conditions for both peat harvesting and reforestation. In the early 1900s, the Karinmossen became a profitable centre for peat extraction and a factory was built in 1938 to crumb, dry and press the hand-cut peat into bales. A five-kilometre-long cable car line carried the baled peat to the railway station at Gysinge and then loaded it onto freight wagons before it was sent across the country. The peat moss factory operated until 1971.[29]

Amongst the managed coniferous forests, mires and lakes in this part of Sweden, Mackmyra's *torv* is harvested by machine from the old salty lake bottom before being pressed and dried into sausages and sent to the distillery. The whisky makers especially value the top layer of white mossy peat. This is less dense, well-structured and ideal for producing smoke that when combined with a little combusted local juniper wood in the smokery produces a uniquely flavoured peated malt. The owners wanted to produce something distinctive and Swedish, and they have done just that.

A New Age of Peat Terroir?

It's 5pm at Kilchoman Distillery on Islay and Derek Scott has just finished his day shift. I jump into his pickup and we head off down the dusty track towards Loch Gorm. Derek asks me if I am a birder as he points out a hen harrier scouting low in a buffeting westerly blowing across the open fields of barley. Ten minutes later, we are looking at a beautifully fashioned, two-feet-wide, four-feet-deep peat bank with three dead-straight cutting levels: top, middle and bottom. Clearly the work of a craftsman. Derek is cutting peats for the Kilchoman peat kiln on the Rhinns of Islay. He starts by picking up a large turf cutter and by jumping down on it using his full weight, makes a series of parallel cut marks along a new part of the peat bank. He then carefully removes the turfs with a spade and places them on the wet bare peat beneath his feet, creating a new vegetated surface from which peat will regrow. The vegetation is still intact, the turf light-brown, fibrous, rich in tough grasses and full of roots. For the top peat, he cuts horizontally along the direction of the bank to create the base of the next layer and then using a peat iron makes downward diagonal cuts, removing each fresh slice and lifting it onto the surface of the bank above. After cutting about ten long rectangular peats, Derek jumps out of the bank and gets hold of a dangerous-looking pitchfork, moving the peats a few metres back from the bank to dry. He lines them up side by side in ordered columns. He then returns to his bank and in the same way cuts and removes the middle layer of peats and then the bottom layer. As he digs deeper, the peat becomes darker as the visible fibres all but disappear. Small twigs of silver birch begin to make an appearance amongst the black homogeneous organic matter. The peats become easier to cut the deeper he goes, but denser and heavier, making them more difficult to throw. He has now cut down to the water-filled trench at the base of the peat bank. This hides geological treasure: the black peat suddenly changes to a creamy white-grey colour – the clays and fine sands of the former Loch Gorm, a huge salty lagoon now cut off from the sea and shrunken in size.

Derek hands me the peat spade and tells me to get on with it, and I then proceed to trash his exemplary Islay peat bank. He quickly calls a halt to proceedings before I can do too much damage to his peat bank, or his reputation, and we take a drink. His peat-cutters' tools are carefully laid out next to us; the asymmetric turf cutter like some giant meat cleaver, the peat iron, which in Islay has a handle traditionally made with cow horn; a conventional flat garden spade and the long-handled, sharp, evil-looking pitch-fork. Later he tells me that I showed 'great potential' as a peat cutter...

Between us, Saligo Bay and the open Atlantic lies Loch Gorm – and if you are a drinker of Kilchoman single malt whisky, you will know these names. We are standing just 15m above current sea level, about 4km (2.5 miles) inland from the ocean on a community of bog-forming plants that include tough grasses, sedges, reeds and mosses. There is also bog cotton and heather and the odd solitary scrub willow tree. Derek is cutting true Islay maritime peat. Standing here 10,000 years ago, we would have been submerged beneath the waves as the oceans swelled after the melting of the last glaciers and ice sheets. Looking around the Rhinns, it's easy to see the old shoreline now 30m above current sea level, including the raised beaches and old fossil sea cliffs that overlook the modern distillery at Rockside Farm. Back then, large parts of Islay would have been under water and the Rhinns severed in half, with Bruichladdich Distillery and Port Charlotte cut off from the rest of the island. When the land bounced back and sea levels receded, peat started to form in the low-lying salt marshes left behind. As we pause our conversation and the wind drops, it's just possible to hear the sea. Derek's peats will dry quickly in the warm air blowing off the Atlantic, and after a few days in the sun and wind a hard, waterproof crust will form and they will start to crack. The turfs will then be turned, a second crust allowed to form, and then they will be stacked into small raised ricks to complete the drying in the Islay air.

Derek has an unusual CV for an Islay peat cutter. He is a Kelso man from the Scottish Borders and a former Royal Navy helicopter engineer with two tours of duty in the Antarctic. His day job is master maltster at Kilchoman, part-time warehouseman and relief stillman. He has no training or qualifications in the fine art of making whisky and has simply learnt from others by osmosis. He never cut peat before he came to Islay, and in the same way he has watched, listened and learnt from the locals. 'When I first cleaned up the old bank and started cutting peat, my neighbours would regularly stop by offering advice from the road. Now they don't.' He tells me Bowmore Distillery once cast their peat from the Loch Gorm area, but most of the banks have long been abandoned. He is rightly proud of his work – a peat bank crafted with the precision of an aircraft engineer with each turf – top, middle and bottom – lined up systematically.

Derek 'loves it on Islay' and in his free time he tells me he stalks, fishes Islay's lochs, watches birds, and is building his own home. He is environmentally aware and we talk about the famous whisky versus geese saga of Islay's past when conservationists met the whisky industry head on in the Battle of Duich Moss. In a battle that the conservationists and the geese ostensibly came out on top, it is strange to reflect that even now a small quota of Barnacle geese still get shot, under licence, by the locals.

Kilchoman is the oldest of Islay's new wave of distilleries. It is making some of its whisky the old way – local barley, floor maltings and hand-cut peat. I ask Derek to tell me what will happen next to his peats:

We usually peat the malt for around ten hours burning about three to four bags of peat each batch, along with some dross or peat chaff to cool the fire down. Each bag contains on average about twelve good peats. With a five to six-day cycle, our malting schedule carries on throughout the year. That is approximately sixty-five fires per year, which results in just over 3,000 peats burned. It sounds a lot, but I would say that one cut of my bank is more than enough to supply the distillery for a year.

Cool-smoke kilning at Kilchoman produces a malt with a phenol content of 20ppm, with peat running like a thread through all of Kilchoman's core range of single malts. Fittingly, the peatiest expression is called Loch Gorm.

* * *

I leave Islay and travel further along the Atlantic fringes of Scotland to the Isle of Lewis, in search of more new-age peat terroir. Driving south from the peat-rich flatlands of the north, the road ducks and dives along the coast, in and out of bays and inland across the rocky Lewisian landscape dotted with landlocked lochans. The Stones of Callanish appear then disappear in the rain and the road turns into a roller coaster switch-back of blind summits and fast descents. It deteriorates further as we get closer to our goal.

Most people drive straight past Abhainn Dearg without realising it; it is like no other distillery I have visited. There is no sign, no pagoda roof and looks more like the buildings of a fish farm, which it once was. I met Gaelic speakers on the Isle of Lewis who didn't even know it existed. I visited with Paddy on a rain swept day in October and he was more than happy to seek sanctuary within this odd collection of buildings. It was one of those days when the Atlantic seems to merge with the rocky west coast of the island, a day when sea spray and sea mist become inseparable. One of those days when low cloud parts fleetingly to reveal the tantalising glimpse of glaciated inland mountains of Lewisian Gneiss.

The distillery is owned and run by crofter and whisky maker Marco Tayburn, a well-known figure on the island. The distillery gets its name and its water from the Red River, a peaty orange-brown stream that cascades off the mountains past the distillery as it makes a quick return to the Atlantic. Water has a fast turnaround time in these parts and that makes this distillery's supply particularly pure and untainted.

Everything about Marco's distillery is unique. Modelled on the design of the island's illicit distillers, its two locally built copper stills look more like a pair of salvaged rivetted ships boilers. They are connected to long, thin lyne arms that drop down into a pair of wooden worm tubs. The hand-built maltings is special. Looking like a large aluminium animal feeding trough, today it contains malt that is slowly drying from the heat of a peat-fired, cast-iron stove. Peating of the malt takes place when the door to the outside is closed, creating a smoke-filled shed of peat reek. Peat is cut locally from a nearby coastal village and brought in from peat banks near Stornoway along with local barley grown from Marco's own farm. Apart from a hydrometer and thermometer in the spirit safe, there was not a scientific instrument in sight. The 'bottling plant' is a table with two demijohns containing liquids of contrasting colour, a few bottles and corks, a filter funnel, measuring cylinder and a 1978 moth-eared copy of *Practical Alcoholic-Strength Tables*.

The whisky made at Abhainn Dearg has been described by some as 'hot and fiery', maybe not unlike that made on these islands all those years ago. I would agree, and the 10-year-old hits the back of your throat with a thud, but a light-coloured whisky I tasted matured in a madeira cask was excellent – a nice balance between the heavy peat and the sweetness derived from a good cask. At the time of our visit, the distillery had been silent for eighteen months – one of the wooden worm tubs had become so dry the staves had shrunk and the stillroom had a sad, melancholy feel about it. I hope this is a short interlude. Devoid of signs, branding, marketing and well-scripted tour guides, there is so much to admire about the philosophy of making peated whisky here on the rugged Atlantic coast of Lewis. Without using the word 'craft', it is the antithesis of the corporate whisky world.

Convinced that we are now in a new age of whisky peat terroir, I called a halt to my wanderings along the Atlantic coastline, but away from the bogs of Sweden and the islands of Scotland, there is much evidence of a rebirth of the bond that links local peat to whisk(e)y. Ireland has not forgotten its distant links to peated whiskey and the boglands of Donegal were a place once renowned for illicit stills. As the warm humid air of the Gulf Steam hits the Donegal cliffs, it creates some of the best conditions for peat growth in the world. Soil surveyors who mapped this part of Ireland in the 1970s found that peat soils covered an astonishing 62 per cent of the land surface of West Donegal.[30] Peat or 'turf' has always been part of the economic and social fabric of this part of Ireland and in the past, it was cut and strip-mined as a fuel that became known as the 'Coal of the Country'. Over 200 years ago, it would have been used in large quantities to make whiskey and it is appropriate that just inland from some of the tallest cliffs in western Europe, Sliabh Liag Distillers have opened a new distillery to make an Irish peated whiskey using barley grown in County Meath and Donegal-cut peat from a bog called *Mín na bhFachraín*. It started distillation in early 2020 and will produce the first legal Donegal whiskey since the Burt Distillery closed in 1841. James Doherty, founder and distillery manager of Sliabh Liag Distillers, has taken his family back to its recent past – his grandfather was making 'a smoky, double-distilled spirit under the authorities' radar on the hills up the glen in Kilcar'.

Whether you are a fan of the concept or not, terroir is a word that is strongly associated with Bruichladdich Distillery on Islay. In 2017, its French owners Rémy Cointreau bought further into the concept by purchasing the Seattle-based Westland Distillery and Domaine des Hautes-Glaces, a maker of organic single malt whisky located in the French Alps. Westland Distillery in the Pacific Northwest employ what I would call a 'whole ecosystem' approach to their whiskey making. Refreshingly transparent about their methods, they use locally grown barley and indigenous trees such as *Quercus garryana* (Garry oak) to make casks.

Westland began producing whiskey in 2010 and after originally importing peated malt from Scotland, decided in 2016 to start using small amounts of local peat from the cedar and pine-lined shores of Oyster Bay, one of the innermost arms of Puget Sound. Their locally peated whiskey, which is due for release in 2023, will be called Solum, a Latin word meaning 'soil'. The marketing department clearly loves peat, 'Primordial Earth – Forever Compressed – The Secrets of Ages – Released by Flame', they have created their own *Peat and Whiskey* mini-movie.

Washington State is home to 50,0000 acres of peat bog[31] and Westland's peat is excavated from a coastal, undrained bog that contains preserved beaver dams in its stratigraphy and a layer of identifiable volcanic ash from an ancient eruption in nearby Oregon.[32,33] Master distiller Matt Hoffman describes the developing aromas of his in-progress Solum single malt whiskey as distinctive, with 'notes of mezcal, herbaceous leaves and roots, green and earthy, spicy and medicinal'. This reflects the surface vegetation of the bog – quite different from what you will find in Scotland or Ireland. 'Rip off some of the leaves and crush them in your fingers. You get citrus and rosemary and lavender all rolled into one.' There is much more in Westland's peat than *Sphagnum*; this is woody, shrubby peat, with aromatic Labrador tea, a fragrant, citrusy relative of the rhododendron, and cranberry. It is herbal and fruity, maybe not unlike the floral peat used by Highland Park cut from Hobbister Hill on mainland Orkney. In another first for Washington State, the Skagit Valley Maltings have built a bespoke peat smoker that is hooked up to a rotating drum containing the drying malt. This is the first place peated malt has ever been made in the USA and for that reason, the whisky world is excited to taste Westland's Solum single malt.

But there is already a hint of what may lie ahead from Jason Parker, co-founder and distiller of Copperworks, the sister distillery of Westland, which shares the same peat source and maltings. He describes smoke from the local peat as 'mild and delicious, unlike that produced from Scottish peat'. Not only are the peat-forming plants distinct, but the process of malting is unique:

The resulting whiskey from our first peated batch is mild and fruity, rather than aggressively smoky. In fact, the aroma doesn't even suggest smoke, and it's barely detectable on the palette. However, it has a delicious, dried fruit and barbecued brisket flavour, followed by a driftwood fire smokiness that dominates a long finish. While lovers of highly-peated Scotch tend to wish for more peat character, we find the vast majority of Americans prefer this more gentle level of peat. Quite often, our first batch will have been the first peated whiskey they have ever purchased.

A global journey of peat terroir takes you across the Pacific Ocean to Hokkaido, an island that contains 90 per cent of Japan's peat bogs. Only 2,000km^2 remain of the island's pristine peat and much is found in the lower reaches of the Ishikiri river valley. Japan's peatlands have been modified and reclaimed by humans for rice cultivation and during World War II when Japan became short of fuel, peat became a vital source of energy. Hokkaido has been called Japan's Islay and when Masataka Taketsuru returned home from Scotland in 1920, he eventually secured the financial backing to go it alone and make a Scottish-style single malt whisky on Hokkaido. He chose a site facing the Sea of Japan and Yoichi, Japan's first whisky distillery, opened in 1934. The climate of this northern island was cold and damp and it reminded him of Scotland. His coal-fired stills were based on those at Longmorn Distillery, where he had spent time, and he set out to make a whisky with a heavy, burnt peaty profile. Nowadays Yoichi, which is part of the Nikka company, imports peated malt from Scotland, although the local peat is still used for demonstration purposes. On the other side of the island, Akkeshi Distillery started making peated whisky in 2016 with an aspiration to make an Islay-style heavily peated whisky using local peat.[34] Clearly buying into the philosophy of Masataka Takesuru, they produced their first peated malt in 2018. Elsewhere in Japan, the cult Chichibu Distillery also has long-term plans to use local peat from the Saitama area. Currently, they import their peat from north-east Scotland and malt it to around 50ppm, a similar level to that used at Ardbeg. The ultimate goal for Akkeshi and Chichibu is to produce a whisky made from Japan's local resources and like the new peated whiskey from Washington State, they fully expect it to be different and distinctive.

Peatlands are of course a global phenomenon and a journey through the world of peat terroir and whisky would not be complete without dipping into the southern hemisphere. It could take you to the Limburners Distillery, where hand-cut peat from the Valley of the Giants in Western Australia is used to make peated whisky that might contain the aromas of tingle, mallee and gum trees. But more likely it will lead you to the famous, trailblazing Lark Distillery where they use hand-cut peat cut from the Brown Marsh Bog, once Tasmania's last working peat mine. The marsh, a small wetland surrounded by bush and gumtrees, is located 42° south of the equator, 730m above sea level in the island's central highland plateau. It's five-metre-deep peats were once drained, strip mined and then abandoned to the local sheep to graze what remained of the vegetation. Restoration of this *Sphagnum*- and button grass-rich bog is now well underway and actively supported by Lark, which takes just six pick-up truckloads of peat a year, allowing the bog to renew. Brown Marsh Bog peat is burnt in a smoker for six hours and used to create a malt peated to 8–15ppm phenols. I have yet to taste Tasmanian peated whisky, but it is said to be softer than its Scottish equivalent.

I have also not had the pleasure of tasting Belgrove Wholly Shit Rye Whisky, but if you want sustainability, reuse, waste minimisation and are looking for something totally different in the whisky world, Belgrove Distillery, Tasmania is your place and Peter Bignell your man.[35] Farmer, mechanic, millwright, bio-diesel producer and now whisky maker, Peter does it all and much more. He hates waste and takes the ethos of reuse extremely seriously. His sheep are fed the waste grains from his distillery and he has used their dung as a source of smoke and flavour. Some rather dismissingly call it barbecue whiskey.

PEAT AND WHISKY

Our short journey around the world of peat terrior at the beginning of the twenty-first century tells a story of a reconnection between local bogs and local whisky. Many of the new players see it as something distinctive as they seek to carve out a niche in an increasingly busy global whisky market. Others see it as a sensible use of local resources that gives authenticity to their whisky and reduces transport costs as they pursue an agenda towards zero carbon. From a global perspective, peated whisky appears to be at the start of a fascinating journey that embraces the geographical distribution of peatlands across the world. Karinmossen, Brown Marsh Bog, Stauning's Klosterlund Moss in Denmark, Loch Gorm, Oyster Bay, Mín na bhFachraín all occur in different, botanically diverse climatic regions. What will the whisky world make of these new spirits; will we be able to detect nuances of aroma and flavour that separate the northern and southern hemisphere whiskies?

It is early days, but to get a view on what might lie on the horizon of global peat terroir, I asked Dave Broom, someone who travels the world exploring, musing, tasting and writing about whisky, for his thoughts:

We have to wait to try Westland's Solum, but Copperworks in Seattle have released a locally peated whiskey with a smoke element that was quite earthy with touches of hickory barbeque and campfire, although the ageing in virgin oak might have an influence. I found Lark fragrant and lifted, almost with a eucalyptus aroma, some mossiness and oils. Belgrove is more burnt, Limburners in Western Australia more fragrant. Mackmyra use Swedish peat for its Svensk Rök and the addition of juniper at the kilning stage for me creates the signature, but there is a tobacco element in there as well. Denmark's Stauning uses local Jutland peat and added heather that gives a gentle mix of earthiness, cacao, bonfire wood smoke and a dried floral edge. Global peat terroir is going to be a fascinating area to explore.

It becomes even more exciting when you consider how the peat is being burnt and the barley malted – shipping containers, bespoke smokeries, the addition of juniper wood and heather.

In its broadest context, peated whisky is of course a global phenomenon and has been for some time. Across the world, it is made in different countries that have developed their own styles and flavours: Amrut from India speaks to me of smoky toffee apples on a childhood bonfire night; Yoichi from the peat-rich island of Hokkaido – salty, fruity with the dry smoke of our old home coal fire. Ashok Chokalingham, former brand ambassador and now head of distillation at Amrut, told me that although they have been making peated whisky since 1995 for its main markets in Europe, Pacific Asia and the Americas, Indians are now beginning to develop a taste for peated single malts. They produce a sizeable 250,000 litres of peated new-make spirit each year and have always used imported peated malt from Bairds in Inverness, using New Pitsligo peat. A newspaper article in 1977 voiced concerns about the rise of Japanese whisky, its quality and the threat it posed to native Scotch. But what really hurt was the import of one of Scotch's most prized natural assets: 'Without the Scottish peat, the Japanese would be lost'.[36] As it turned out, not the most prescient of comments, but a reminder that peated blends or single malts made around the world use almost exclusively either peated malted barley or simply peat imported from Scotland.

So, whether you buy into the concept of peat and whisky terroir or not, distillery peat sheds, so beloved by Alfred Barnard, are beginning to return and local peat is once again being used to make local whisky. Many whisky connoisseurs are convinced they can detect the difference between whisky made from malt infused with Islay peat from that made using peat from the Scottish mainland or cut from a hill on an Orcadian island. I am not so sure and on my journey, as I have learnt more about the process of whisky making and the individuality of distilleries, I have become more convinced that the subtleties of smell and flavour are at the hands of the craftsmen and women who make the stuff. This is terroir in the true definition of the word.

This is an exciting time in the story of peat and whisky and soon enough we will know whether we can make a sensory connection between Westland's Solum smoky single malt and the aromatic peat bogs of the Pacific Northwest, or peated whiskey produced by Sliabh Liag and the Atlantic bogs of west Donegal. While the spiritual home of peated whisky will always be the islands of Scotland's Atlantic coast, my spiritual home of peat, wetland and water lies to the north and east in the UK's largest peatland – the Flow Country of Caithness and Sutherland.

Six

Walking on Ninety Per Cent Water

'There is one advantage in an uncomfortable bed; it induces early rising, and it proved so in the present instance, for we finished our breakfast and resumed our journey by 2am.'
Dr John Rae, Arctic explorer and physician, written on 10th August 1846 on a journey to Repulse Bay.[1]

Thurso, April 2013

The ticket inspector smiled when I told him my destination. After closely examining my ticket, he wished me good luck and moved onto the next passenger on the early morning northbound train from Edinburgh. Just beyond Perth, we cross the Highland Line – a former taxation boundary that was once used to levy separate duties on whisky made in the Lowland and Highland Regions. An arbitrary line drawn on an eighteenth-century map by the Excise, it tries to follow the Highland boundary fault, a well-defined geological discontinuity that separates the young, fertile agricultural lands of the Lowlands from the old, hard Highland schists and gneisses with their rough moorlands, mountains and peat. After leaving Inverness later in the day on the old Highland Railway, we steadily meander our way along the north-east coast of Scotland in touching distance of a swathe of distilleries – Glen Ord, Teaninich, Dalmore, Invergordon, Glenmorangie, Balblair and finally Clynelish-Brora. Alfred Barnard rode the same railway more than 135 years ago on the final leg of a thirty-six-hour journey from London. He was on his way to catch the early evening steamer to Orkney, visit the distilleries of Highland Park, Scapa and Stromness, and set eyes on the rich cultural history of these far-off isles.

PEAT AND WHISKY

We leave the coast behind and begin climbing slowly uphill into the UK's largest peatland, past old snow fences, summiting at Forsinard Station, before stopping briefly to draw breath. Forsinard, a Norse-Gaelic hybrid word meaning 'high water', marks the point the watershed divides and where the twin sandstone peaks of Ben Griam Beg and Ben Griam Mor rise in the west out of a vast peatland plateau. Peat contains 90 per cent water and we are close to the point at which three of Scotland's great peaty brown salmon rivers, the Helmsdale, Halladale and Thurso, begin their journeys from source to sea. They flow south, north and east from the heart of this great mass of bogs, sepia-coloured pools and lochs elevated 100–200m above sea level.

Forsinard has its own small place in the story of peat and whisky. After the Highland Railway opened in 1874, the Duke of Sutherland built a small peat factory here amongst the bogs to experiment with making peat charcoal for Scotland's iron smelters.[2] The peat was baked in a small furnace to drive off water and sulphur, but apart from providing overnight shelter for twenty stranded railway workers in a huge snowstorm,[3] the factory was a commercial flop. The new railway opened up this part of the Highlands and on it travelled livestock and goods, as well as passengers. Special 'sheep trains' ran south each autumn as stock was moved to market in an annual exodus from the grazing lands in the northern glens. It also carried peat north to the town of Wick and south to distilleries including Clynelish in the coastal town of Brora. Peat is not a word you would associate with the modern taste profile of Glenmorangie single malt – light, floral, sweet and citrusy are more likely – but at the turn of the twentieth century, the distillery was cutting 'several thousand tonnes' of peat each season from the mosses close to Forsinard station.[4] The distillery's maltmen, accompanied by a seasonal workforce, would travel north in late May or early June and spend two weeks away from home. After a day working at the peat banks, they lodged with local shepherds and estate workers. Later in the summer, the Glenmorangie Distillery Company would hire wagons from the railway and the returning workers would load the dried peats onto trains at Forsinard Station for the short downhill journey south to Tain.[5] The peat trains from Sutherland continued until after World War I when the Forsinard peat banks began to run out and the company found new suppliers in Orkney and closer to home.

We are soon off again from Forsinard Station and the train races downhill towards the coast for the remaining forty minutes of its journey. We speed through Altnabreac Station without stopping. In 1882, a newspaper correspondent on a grand tour of the north did stop here and wrote 'Arrived here five minutes before time... observed one small boy with a windcock, several hurleys, and all around peat, peat, peat. Not a single passenger came in or got out.'[6] It is still a remote community in amongst the peatlands, a place of wild weather, and somewhere that I will return to soon. Its deep peat bogs have at various times attracted interest from scientists, power generating companies and the nuclear industry.

Nine hours after leaving Edinburgh, I step down with my fellow travellers from the two-car diesel train onto the station's only platform. We have arrived at the very northern edge of Scotland, the end of the line. '******* to Thurso' reads the Scotrail sign above the platform. Thurso had temporarily lost its 'Welcome'. Located 58°35' north of the equator, the sun will set today at 8.30pm and rise again at 6am. Barnard had come this way later in the year than I, in the long summer months, but the days are lengthening quickly now, and spring has arrived. As I walk down Princes Street to the sea, high up in the tall beech trees the rooks are busy and noisily going about their business. West Beach is unusually calm with no brightly coloured surfers to draw the eye out in the bay. Today's waves are little more than ripples, presenting no challenge to the local boarders. In the bright, clear April air, the rusty brown Devonian sandstone cliffs of Dunnet Head, and further in the distance, mainland Orkney, look much closer than they really are.

These are exciting times for whisky making in these parts. Almost three months ago, after a hiatus of 150 years, the new Wolfburn Distillery fired into life. The original distillery built in 1821 was once the largest producer of single malt whisky in Caithness. Using locally harvested peat and barley, it distilled what its first owner William Smith liked to call 'Thurso Aqua Vitae'. Appropriately, it was on Burns Night that new-make spirit began to flow again in Thurso and Wolfburn regained the crown of Scotland's most northern mainland distillery. The 'Wickers' down the road at Pulteney Distillery had competition in Caithness once more.

From Thurso, I will embark on a journey into the UK's largest peatland, but it is worth pausing for a moment to contemplate both the size and international importance of the area we know as the Flow Country. 1: it is the largest blanket bog in Europe, covering a total area of just under 6,000km². 2: this interconnected area of bog in the north of Scotland contains a quarter of the UK's peatland. 3: around 400 megatonnes of organic carbon are stored here at the earth's surface in a two to three-metre-thick layer; a fragile skin of peat covering rock and glacial debris below. To some, the Flow Country represented a considerable resource to be exploited for fuel, or a 'Bad Land', a vast area of barren and unused ground waiting to be improved, cultivated and planted. Alfred Wainwright apparently saw no beauty in the Caithness landscape: 'There is nothing for the lover of countryside here,' he wrote.[7] To others, this remote, sparsely populated tableland of peat was an opportunity to dispose of significant quantities of long-lived hot radioactive waste – out of sight and out of mind for future generations. Fortunately, we now live in more enlightened times when the peatlands of the Flow Country are valued in a different way; they are protected and their meaningful exploitation by humans largely at an end. But even in 1955, Frank Fraser Darling, one of the most respected ecologists of his time, called these *Sphagnum*-rich landscapes 'wet desert'.[8] Deserts to some, maybe, but these deep, wet, cold peatlands are places of archaeological treasure – hot spots of preservation. One of the most recent finds, grasped from the teeth of a crofter's digger, were the remains of a wolf cub that died 2,000 years ago.[9]

There is quite simply no better place in the UK to immerse yourself in everything about peat and explore its story and the changes brought about by humans that continue to this day. Scientists have now arrived to make measurements, unearth evidence about the past, collect new data to help understand and predict the future. Like many parts of the world, the climate in the north of Scotland is changing and the peatlands of the Flow Country will not be immune to the impact of these changes. Scientists work hard to tease apart the 'signal and the noise', as Nate Silver put it so succinctly in his book.[10] The signals are the long-term changes caused by climate change such as warmer winters, species abundance or an increased frequency in flood events or droughts. The noise is the 'static' or 'white noise' of natural variability often associated with short-term seasonal fluctuations in weather. Scientists are now much better equipped to unravel the signals from the background noise, but linking cause and effect is always our greatest challenge.

As a working scientist, I have come to Thurso time and time again over the last two decades, and it is a place I have grown to like and know better than I once did. In recent years, it has become a hub for peatland researchers centred around its Environment Research Institute. Regular late-night peatland research chat with scientists from all over the world was usually accompanied by a dram at the Commercial Hotel. Always a lively place, Thurso was never going to be a temperance town like its neighbour Wick, which was dry for twenty-five years, almost twice as long as the period of American prohibition. Establishments like The Comm were more than happy to receive the locals from down the road. Scientists have never been very attached to the concept of temperance anyway.

In the coming days, I plan to walk south from Thurso across the flat Caithness farmlands and then west, meandering my way through this northern peat-rich landscape until the top half of Scotland spits me out at Kylesku on the Atlantic coast. Caithness is a place of whisky making, but I am heading towards a whisky desert, the rocky ancient crag and lochan landscapes of Sutherland and the wild, wet, infertile west. I had spent twenty years researching the peatlands in this part of the world, now it was time to journey; to walk, look, see, listen, learn and escape. The only problem was the weather. As I sat in my office watching the daily weather forecasts unfold, the north of Scotland had been blessed with perfect weather for the last month with dry, cold, winds from the east assisting the westbound walker – according to the forecasts, that was all about to change.

Although the Romans never reached these northern lands, preferring the relative security behind the Antonine Wall further south, the road that links Thurso to Halkirk feels like it was built by them. It heads dead south and straight, keeping the lazy meandering Thurso River in touching distance to the east. Field boundaries of weathered Caithness flagstone intersect the road at right angles, the individual slabs of rock placed vertically end on end and dug deep into soil long ago. Spring fields, still struggling to shed the colours of winter, are full of twin lambs and their protective mums. They nibble away, low down seeking out new and nutritious shoots of bright green grass. Where it has survived the attentions of the

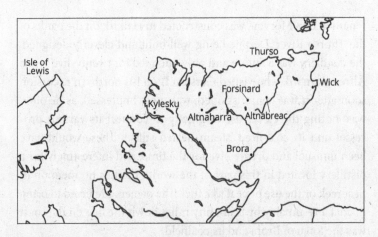

local grazers, isolated clumps of brilliant, bright-yellow, coco-nut-scented gorse are in full bloom. Clouds are beginning to gather, gusts of wind out of the south-west smack you in the face, and last year's wool snagged on barbed wire is becoming increasingly agitated. A rainbow out to sea over Thurso Bay completes the signs of spring on the north coast of Scotland.

It is hard to imagine picturesque Halkirk, Scotland's first purposefully designed new town, with its delineated housing plots, poorhouse, neatly kept gardens, village green, church and hotel, as a whisky town. Gerston Distillery pre-dated Halkirk, and like so much of Scotland's rural whisky production, was rooted in an illicit past; local bere barley grew well this far north, peat fuel and fresh water were never in short supply. It was first registered as a distillery by Francis Swanston in 1795 and although it never reached the production levels of neighbouring Wolfburn, it was noted for the quality of its whisky. It was pulled down in 1882, but during its relatively short life it became a highly respected whisky in London, particularly amongst the political classes, and was said to be a drink of choice for Prime Minister Sir Robert Peel.[11] In 1886, a much larger modern distillery, Gerston II, later renamed Ben Morven, was constructed in Halkirk on the banks of the Thurso River. Despite being well-built and cleverly designed, the distillery was a failure and only operated for twenty-five years. Alfred Barnard, who visited Gerston II on his northern excursion soon after it was commissioned, was most impressed, as he often was, noting the clever use of gravity to connect its various processes and its coal-fired, steam-heated stills.[12] These would have been unusual and progressive for the time, and interestingly for a distillery located in this part of the world, there is no mention of peat reek or the use of peat as a fuel. The owners preferred to bring in coal, presumably by train; fifty miles down the line on the coast was the town of Brora and its coalfield.

In 1803, Halkirk was chosen as the site for a new town because of its location as a major crossing point of the Thurso River, one of the main freshwater arteries that drain the heart of the Flow Country. It is one of Scotland's best Atlantic salmon rivers, with a catch of around 1,000 fish each year, and has excellent stocks of brown trout. It is also a highly coloured river carrying peat-rich water from inland areas to the north coast and out into Thurso Bay. Each year, this single river, with a catchment area of over 400km², transports almost 4,000 tonnes of carbon to the sea, equivalent to the annual loss of a two-metre-deep peat bog measuring 100x100m.[13] To put this into perspective, the Scotch whisky industry uses between 6–7,000 tonnes of peat a year. The peats of the Flow Country are naturally 'dissolving' and along with small peaty particles, the carbon leaches away continuously from the bogs. Eventually they find their way into the Thurso River and its tributaries as it snakes its way across the Flow Country on its journey from source to sea. The river is acting like a continuous conveyor of organic carbon and in spate becomes overloaded with peat-rich water. When it reaches the sea, it sometimes produces a distinct brown plume that is clearly visible from the air as the river discharges its load of Flow Country peat into the clear ocean.

Just outside Halkirk, I cross the same railway that I had travelled along earlier in the day and head on south past Berry Croft, the home of Caithness Peat, the local peat extraction company. On the banks of the Thurso River is the working mill of Westerdale and a spectacular water-sculpted sandstone platform that shows how hard the river once had to work to breach this formidable barrier. I have now reached the edge of the Flow Country and my map begins to tell a story of approaching bog. Moss of Halkirk, Yellow Moss, Bloody Moss and Dale Moss all pass close by, and tellingly the blue lines on the map change from wandering snake-like meanders to purposeful interlocking sets of parallel lines – a man-made symmetry of drains.

Caithness peat has a history of use in whisky production, although unlike the distilleries of Islay, has not found its way into the DNA of locally made whisky. Things would have been different in the past, but the house style of modern Pulteney, Wolfburn, Clynelish and Balblair is now a lighter, fruitier new-make spirit, with only rare excursions into peated expressions. In 2013, Caithness Peat was licensed to extract peat from several hundred acres of the Causeymire Moss for fuel and malting barley for whisky production. It is a throwback to a time in the 1980s and 1990s when large-scale mining of Caithness peat was commonplace and further expansion and new enterprises were being planned. At its core was the Highland Peat Company, then the largest peat extraction company in the UK, and plans were being floated for a barbecue charcoal briquette factory and a peat-processing plant in Wick. I well remember the peat-mining operation in the 1980s at Causeymire – its scale, the strange machines and the huge piles of peat sausages disgorged by harvesters that each year steadily removed layer upon layer of peat. Although the modern operation at Causeymire is now much smaller and the site largely abandoned, the scars of the past are still evident, with little or no effort to reinstate and restart the natural process of peat formation.

In a strange twist of fate, a modern wind farm now occupies the site. Construction began in 2003 and on my many journeys to the Flow Country at the time, I recall how the turbine count steadily increased. Each time I drove past, at the end of a newly constructed access road there was yet another 60m-high white turbine, ready and waiting to start harvesting energy from the wind. The Causeymire cluster of turbines now generates up to 48.3 megawatts of electricity, sufficient to power around 30,000 homes, or a city the size of Inverness.

That evening, with the wind turbines at Causeymire spinning away in the distance, I walked past the old hunting lodge at Strathmore and at the edge of a forest of Lodgepole pine, decided to make camp on boggy tussock grass. This was the first of many forests planted on peat I would travel through in the Flow Country that with time would grow to become a major battleground between the forestry industry and conservationists. I was now close to Loch More, one of the largest areas of open freshwater in Caithness. For a short moment after dusk, the air became full of the noise of wing beats and the quick-fire repetitive squawks of Icelandic greylag geese, returning to their overnight roost after a day grazing on slim pickings in the farm fields of Caithness. I settled down for the night and listened to the sounds of wind in the pines.

* * *

War visited Loch More on the night of 5th June 1940:

Police Constable George Sinclair of Halkirk recorded that at 11.45 on the night of 5th June, a bomb had been dropped on Loch More. It had been reported at 12.05 by John Mackay, gamekeeper at Strathmore, saying 'that about fifteen minutes previously an aeroplane, the noise of the engine of which sounded similar to that of a German aeroplane (something like 'hoo – hoo – hoo'), had circled above the lodge and had then passed over Loch More, into which a bomb, which exploded with terrific concussion, was dropped'. Fortunately, no one was hurt, though two men had been fishing in nearby Loch Beg at the time.

The German plane had unloaded its heavy bomb after failing to find a target, prior to a long flight back to base across the North Sea.[14] On 1st July 1940, a short time after this incident and more than two months before the London Blitz, Wick on the Caithness coast became the target of the first daytime bombing raid on mainland Britain. From the early days of the war, the Royal Naval base at Scapa Flow in Orkney had been attacked by the Luftwaffe and night-time raids would occasionally be punctuated by solitary explosions and craters on the mainland as the bombers unloaded unused explosives into Caithness soil before returning to base. The German invasion of Norway now meant that the town was within range of German Junkers Ju88 bombers based in Stavanger. Wick, with its wartime RAF base and harbour, suffered a total of six air raids, the most deadly being the first, when a single bomb from a Ju88 approaching over the sea was dropped on Bank Row in Pulteneytown, killing fifteen people including eight children playing in a late summer afternoon.[15] Officially, 222 high explosives were dropped on Caithness soil during World War II, many failing to explode as the detonator was not triggered on impact with the soft blanket of peat.

The Icelandic greylag geese, serial slow-starters in the annual northerly migration race, were gone by the time I emerged from my tent just after dawn. Maybe they were already *en route* to

Iceland, impatient to catch up with the other migrants already well on their way. It had been a restless night and soon I discovered that I had parked my tent on a deer path, and by the number of ticks crawling purposefully around in the vegetation, a place where deer also regularly chose to rest for the night. I made a mental note to check that evening that I had not become a new host.

By mid-morning, Loch More had been transformed into a raging white waterscape. Strong winds were blowing down the loch and generating short waves that wash over the track and lick my boots. The atmosphere is crystal clear. Clouds move fast in a big sky above the wave-torn, dark blue water. The loch fills a large and shallow flat-bottomed peaty basin, and at times of drought the water level drops far enough to expose large deposits of black peat-rich sediment brought down by rivers and streams from the surrounding peatlands. It is so shallow that on occasions it completely freezes over, as it did in the winter of 2010. In the same year, elevated phosphorus levels were found in the loch, a result of pollution from past fertiliser applications to the surrounding man-made forests. The loch also marks an important geological transition in my east to west journey as I leave behind the young porous Devonian sandstones that cover much of Caithness and set foot on the older metamorphic rocks of the Moine Schists. Covering large tracts of the Scottish Highlands, they are named after the Gaelic *A' Mhòine*, or 'the Peat Bog', a vast peat moss in the far north-west of Scotland. As I travel further west into the Moine landscape, it will become progressively wetter, the peats deeper, and its pools and lochans places to be avoided or navigated around.

To the west, the twin peaks of the Ben Griams are now coming into view, and dead south across a Caithness plain dotted with long-abandoned farmsteads are Morven, Smean, Maiden Pap and Scaraben, which form a familiar skyline in the distance. At 706m, Morven is the highest peak in Caithness and gave its name to the old Ben Morven Distillery. Its summit was high enough to poke through the 500m-thick ice sheet that once covered Caithness, and 20,000 years ago it would have been a lifeless, harsh place of frost and sand-blasted rock.[16] The Inuit call these places *nuna-taks*, a cold, dry and windy desert island that escaped the erosive power of the sea of ice moving steadily across the plain below. After the ice disappeared around 10,000 years ago, peat began to form on the flat and wet Caithness plain. With the weight of the ice removed, the land surface rebounded and gradually rose up, now lying 100–200m above current sea level. In 2017, Wolfburn Distillery added Morven to their core range of whiskies; on its label is the characteristic double-peaked profile of the mountain's summit. It is lightly peated, maybe a historical nod to the old nineteenth-century distillery that used peat to fire its stills.

Leaving Loch More behind, the path enters the comparative shelter of a large blanket of coniferous plantations. I have now entered the forests that caused so much controversy in the 1980s, although the degradation of the Flow Country's peat started much earlier.[17] The name 'Flow' is derived from the Old Norse word *floi* meaning 'marshy ground', so peatland had to be drained before it could be planted and 'improved'. This started with the arrival of the Cuthbertson plough in the 1940s and 50s. Invented by James Cuthbertson, a native of Biggar in the Scottish Borders, this huge plough was pulled by a crawler tractor with a single tooth that would rip open the peat to construct a deep drainage channel and throw up a turf on to which a tree seedling could be planted. The most intensive period of peatland afforestation took place in the 1970s and 80s, driven by tax incentives from the UK government. Terry Wogan, Nick Faldo, Cliff Richard, Alex Higgins and Dame Shirley Porter were among those who lined up to invest their celebrity money into planting non-native, exotic Lodgepole pine and Sitka spruce forests. The speed of planting was staggering, and by 1988 it was estimated that 25 per cent of the peatlands had been drained and planted. The key to turning the tide was research that linked afforestation and wetland destruction to the decline of the nesting wader population. A moratorium on tree planting was announced and this provided the platform for the future conservation and restoration work that followed. Through a series of major land purchases, the RSPB gradually became one of the largest landowners in Caithness and has been at the forefront of an extensive peatland restoration programme. Since bogs can shrink by as much as 25 per cent when drained, blocking those drains and removing the trees were the first crucial steps in returning the Flow Country to its former state of wetness and bogginess.

The well-made forestry tracks, built high and dry above the peatlands, strong and wide enough to take timber lorries, are excellent underfoot. The surrounding trees provide shelter from the wind and I make steady progress westwards towards Altnabreac. Either side of the track are deep, water-filled drainage ditches exposing peat containing large black-and-white striped boulders of Moine Schist. Poking out from the sides of the drainage ditches are pieces of ancient wood, mostly pine. Occasionally, the sound of the railway reaches me across the peatland forests, a steady two-car rattle, which arrives in short waves of sound that become louder before gradually fading away as the train trundles northwards towards the coast. Sound travels great distances across the Flow Country, and on occasions you hear the driver turning up the diesel power as a southbound train pulls harder up the gradient towards Forsinard. Four trains each way, each day – a timetable of predictable sound that radiates out across the peatland. I cross the Caithness-Sutherland boundary and as I close in on Altnabreac, the railway reappears. Further along the line, I see a man dressed in oversized bright orange overalls and safety specs walking north with a purposeful stride – a linesman, the first walker I have encountered on my journey so far. Shortly afterwards, I hear that familiar distant rattling sound again, which heralds the appearance of a northbound two-car train, on time and in a hurry to get to the coast.

Altnabreac is a place I have been looking forward to coming back to. In 1988, a discovery was made here that was interesting at the time, but it wasn't until the events of spring 2010 that its significance became apparent.[18] Seventy centimetres below the surface, researchers found a thin layer of peat containing microscopic glassy fragments of a rock called tephra. It was not only the first discovery of Icelandic volcanic ash on mainland Britain, but it contained the unique chemical fingerprint of the Hekla 4 eruption that occurred in 2,310BC, a massive Category Five event of the same magnitude as the 1980 Mount St Helens and 79AD Vesuvius eruptions. It was explosive, long-lived and with a favourable wind led to the transport of small particles from deep within Iceland's oceanic crust to the peatlands of northern Scotland. Over the next 4,000 years, the evidence had become buried in time by the growth of new layers of surface peat.

On 20th March 2010, history repeated itself when another volcano, Eyjafjallajökull, began erupting in southern Iceland and over the next two months caused major disruption to air travel over north-western Europe. The mix of super-heated molten lava and glacial ice in the volcano's central vent resulted in an explosive eruption that sent clouds of dust into the upper atmosphere. The ash landed in many places, including the windscreens of cars in Shetland.

Altnabreac is a place that is central to the story of the Flow Country and a microcosm of the various attempts that have been made to exploit its peatlands. It is easy to imagine a group of planners gathering around a map in the mid-1900s, finding this remote

place in the middle of a seemingly endless bog and pondering how
to put it to better use. It was a settlement at one time, with a school,
a freshwater spring and peat banks for fuel. Out on the bog a few
years ago, I met a former pupil, Patrick Sinclair, who each day
made the long trip from his home at Strathmore Lodge to attend
class at Altnabreac.

In 1879, a local gathering of Altnabreac school pupils for tea
and cake numbered thirty.[19] By 1969, the school had shrunk to
only four pupils and the local education committee was much
vexed with the problem of providing free milk and dinners to a
school with no mains electricity, running water or toilet facilities.[20]
A cow or a goat was even suggested as a solution to the problem.
In 1979, there was just a single pupil left at the primary school and
in 1986, the school closed for good, a story that would be played
out time and time again across many parts of the Highlands. A
white sign next to the school building now reads 'Altnabreac DC,
twinned with Washington DC'. Presumably DC refers to 'District
of Caithness' and not 'District of Columbia'. In 2018, a decision
by the Royal Mail to suspend the delivery of letters and parcels
was met with fury by the four remaining families.[21] The decision
was overturned after a lengthy campaign – Altnabreac had lost its
school, but not its post.

Altnabreac (in Gaelic, 'the stream of the trout') still has a
railway station, but as a passenger you have to make a request
for the train to stop. A sign states 'There are no ticket issuing
facilities at this station' and stopping a train from the platform
involves using rapid hand signals. It is a mystery why the station
should be here at all since the school and the local hunting lodge
were built after the railway arrived. Once, the grand, turreted
Lochdhu Hotel, now a private lodge, would have welcomed
Victorian ladies and gentlemen off the train and driven them
one-and-a-half miles down the road to their rooms. Later in the

evening after supper, the next day's hunting and fishing would be planned over a dram or a glass of port.

With the season over and the sports men and women gone, Altnabreac waited for winter. A newspaper headline, 'The Story of the Snowed-Up Train at Altnabreac', from 1892 captures what a winter can be like in the Flow Country.[22] On board the south-bound Thurso to Inverness train, with ten passengers and five cattle, was a thirty-five-strong team of linesmen. Their job was to clear snow and as the train battled on, with the engine leaving the rails at least once, it was finally stopped in its tracks by huge snow drifts just south of Altnabreac. Three of the linesmen walked west down the line to the Forsinard Hotel, and after a nine-hour trek arrived back at the stranded train with 'oatmeal cakes, biscuits, cheese and some whisky'. The weather improved slightly, and a decision was made for everyone to leave the sanctuary of the train. The party walked 'Indian file' down the railway tracks to Forsinard, stopping to rest and take on water at the linesmen's huts dotted along the way. The party reached the hotel after a three-hour walk and waited for rescue. It came several days later in the form of steam engines, a snowplough and a carriage full of provisions. The passengers eventually reached Inverness, five days late.

Sixty years later, Altnabreac was back in the news for a different reason, 'Power from Peat and Altnabreac to make History' ran the headline.[23] 'The centre of the Government's £500,000 plan for the production of electricity from peat will be Altnabreac, Caithness, where a population of less than fifty cluster round a lonely railway station.' The remoteness of Altnabreac was always a challenge to the community, but no deterrent to those seeking to make power from peat. In the 1950s and 60s, this small community was at the centre of a national drive to exploit Scotland's peat resources for large-scale power production. The Scottish Peat Committee drew up an inventory of peat deposits across the country to fuel locally operated peat-fired power stations, looking at the suitability of the bogs for drainage and ease of extraction, as well as surveying local supplies of cooling water. Altnabreac was an obvious target.

The surveyors found a huge area of exploitable fuel covering 84km^2 with estimated reserves of thirteen megatonnes of wet peat.[24] Now earmarked for milling and fuel briquette production, an experimental power station was built a few miles down the line and a big sign erected: 'North of Scotland Hydro-Electric Board Peat Development Altnabreac Olgrinbeg Power Station. No Admittance Except on Business'. But the early optimism faded, the economics of large-scale power production this far north didn't add up, and the remoteness of Altnabreac once again became the deciding factor. The time and money invested in peat power came to nothing, the project abandoned, and the experimental power plant dismantled.

Altnabreac wasn't forgotten by the energy industry for long and in the late 1970s, it was again in the news: 'Alarm Over N-Waste Plans' screamed the headline.[25] Lying just below the thick cover of surface peat and glacial debris is the Strath Halladale granite, a huge body of igneous rock that was intruded into the Moine Schists around 400 million years ago. It reaches a depth of 4km in the Earth's crust and began to be explored as a safe haven for what had become politically toxic waste. Twenty-seven boreholes were drilled into the granite to test whether it could be used for the deep disposal of the most dangerous high-level hot radioactive waste being produced from nuclear power stations across the UK. The results were positive and Altnabreac appeared on a shortlist of twelve potential sites. In many ways, the site was ideal for the government and the nuclear power industry. It was close to the Dounreay nuclear power station on the north coast of Scotland, it had a connecting railway line, the peatland had little economic value and few people lived there. But the political mood changed, and the plans were shelved.

However, radioactive waste did finally arrive in this remote landscape in spring 1986 after a safety test went catastrophically wrong in reactor No 4 at the Chernobyl nuclear power station in the Ukraine. This led to meltdown, explosion and fire as temperatures in the reactor's uranium core exceeded 1,600°C, and led to the production of 100 tonnes of radioactive lava that eventually cooled to form a material like volcanic glass. At the time, thirty-two people died with many, many more in the following years. It is estimated that fifty tonnes of radioactive material were released into the atmosphere and seven days later after passing over many parts of continental Europe, particles bearing the isotopic signature of Chernobyl began falling onto peatlands in the UK and Ireland. Forty per cent of the landmass of continental Europe was affected by the Chernobyl cloud as fallout reached as far north as Iceland. It can still be detected below the surface in the bogs today, now covered by a new layer of young peat. The forensic value of peat bogs, like those around Altnabreac, to collect, record and preserve the facts through time, was evident once again.

Leaving Altnabreac and its recent history behind, I follow a track that takes me through forest before breaking out onto open ground.

We are now in an area of the Flow Country recently bought by the RSPB, which has cut down trees and blocked drainage channels in a huge program of peatland restoration. Away from the shelter of the forest, the real power of the wind becomes apparent and at one stage I am blown sideways, rucksack and all, off the track and into a peaty ditch. The storm is arriving and inspired by Johnnie Walker I try and 'keep walking'. I struggle to walk in a straight line, some-times forward, sometimes back. It is getting wild and I begin to recall dark stories of tragedy on the open peat moss. I pass a blasted wooden sign with the solitary readable word 'OUT' and a large white arrow, pointing west. I am trying, doing my best. Eventually, I reach a natural depression surrounded by a small area of trees and decide to make camp for the night in a place with some shelter. A dangerous-looking evening sun appears and disappears through veils of fast-moving cloud. Loud bursts of rain began to fall on my small tent. That night, I phoned home and after a consultation involving all the weather reports that could be mustered, I made the decision to leave the Flow Country as quickly as possible. It was now or never – I was reaching the point of no return.

Next morning, that decision looked an even better one. I packed up, quickly stuffing a wet tent into my rucksack and, like travellers in the snow before me, headed off west, shoulder into the wind and rain, towards the next station down the line, Forsinard. The track down to the Halladale River had turned into a water feature and emerging from a forest and coming towards me was a Landrover that I vaguely recognised. It also contained some people that I knew, including the smiling faces of Roxane and Jens, scientists heading out into the field to make measurements whatever the weather, collect samples and download data from their monitor-ing equipment. They were on a mission to unravel the signal from the noise, and it took them a while to realise it was me. These were the first living souls I had met and spoken to for almost three days, and as happens after a period of solitude, the tongue is too loose, speech is garbled and becomes too fast and over-excited; I probably seemed a bit touched. After a short explanation and a bit of chat, we headed off to our respective fates.

Just before the track reaches the road, I pass a large green shed. It symbolises one of the greatest follies to be inflicted on the Flow Country, and there have been a few. What is particularly striking about this shed is its size – it is more aircraft hangar than shed, and at one time was destined to become a facility for grass drying and storage. In the late 1970s, around 30,000 acres of the Forsinard Estate was bought by Basil Baird & Sons, a family of pig farmers from central Scotland.[26] They planned to first drain and then fertilise the peatland – grass would be sowed, cut and dried and sold as animal feed. The large green grass-drying shed, which still stands as a monument to this bizarre and implausible project, was built with a half-million-pound government subsidy. The politicians of the day were keen to support the creation of jobs in remote areas driven by a policy of making barren wasteland productive. The grass-growing enterprise unsurprisingly ran into cash flow problems and like so many others of its time failed, leaving a legacy of environmental damage and broken peatland.[27] Just three years later, the land was put up for sale for offers over £700,000. It was eventually brought in 1986 by Fountain Forestry and became a young conifer planation, another new enterprise heavily incentivised by the government of the day and also doomed to failure. History was repeating itself again in this remote, apparently unwanted place.

I finally reached the relative sanctuary of the main road and walked south to the station. After passing the Forsinard Hotel and receiving a welcome cup of tea at the damp but warm RSPB Office, I climbed up onto the station platform and waited. It was not long before I heard the familiar distant sound of a two-car diesel train to Inverness and headed home. I had walked from Thurso to Forsinard into the heart of the Flow Country and planned to come back sometime soon.

Forsinard, March 2019

'This is Kinbrace. This train is for Wick. The next stop is Forsinard.' Shortly after the on-board announcement, we pass the familiar bleached and broken snow fences and the abandoned Glenmorangie peat banks and step down onto the platform at Forsinard, the nineteenth stop on the line. I am back on the railway tracks that once brought peat south from the bogs around Forsinard and Dounreay to Glenmorangie Distillery, and 100 years later carried radioactive waste away from the nuclear reactors built on Scotland's north coast. Forsinard Station first opened in 1874 to bring the Victorian aristocracy to the hunting lodges of the Flow Country to shoot game and kill salmon and trout. We carry no guns or rods, and after a polite exchange of waves with the guard, shoulder our backpacks and both ourselves and the train head off in separate directions.

We are the only two passengers to leave the train at Forsinard and the station is now quiet. Inverness Station just over three hours ago was anything but, a sea of people, boisterous, inter-weaving stag and hen parties arriving by train early on a Saturday afternoon to fill the bars and nightclubs of the city. Awaiting them was a quiet and pensive group of Trees For Life volunteers, thoughtfully eating their sandwiches from Tupperware lunchboxes. A minibus eventually arrived to take them away for a week of planting native trees species on the peaceful shores of Loch Ness.

I was resuming my 100-mile (160km) journey through the peatlands of northern Scotland that started almost six years ago. This time, I have company in the shape of my brother Bob, a trained paramedic. In a weak moment, he had been coaxed away from his allotment and the creature comforts of north London and persuaded to carry a 20kg backpack across one of the peatiest and wildest landscapes in Europe. Last night, there had been a worrying early sign – Bob was repairing his old boots with duct tape – but if all goes well, seven days of walking lie ahead before we make ocean fall at Kylesku on Scotland's Atlantic seaboard. Seven days ago, Scotland was covered in a blanket of late winter snow; it is still cold, but most of it has now melted and we have just passed the equinox. Spring has officially arrived and the sun will set today at 6.30pm and rise again at 6.15am.

Six years is a long time since my first attempt to walk alone across the peatlands of northern Scotland ended in meteorological mayhem. In between, Wolfburn Distillery has produced its first whisky with the words 'Made in Thurso' proudly displayed on the label, and its core range of expressions – Northland, Aurora, Morven and Landskip – are being exported, sold and drunk in more than twenty-five countries around the world. After Scotland voted to remain a part of the Union in 2014, the UK decided two years later that it wanted to leave the European Union but hasn't yet managed to work out how. The United Kingdom is disunited, and Scotland may need to think again. The North Coast 500 touring route has brought new drivers and riders to the coastline of northern Scotland and new visitors to its distilleries. World Heritage Site status has become a distinct possibility for the Flow Country, and Forsinard has a striking new space-age observation tower to look down on the black peatland pools and look up at the stars in the dark skies above. And Caithness has a new flag; a black, blue and yellow Nordic cross. For the record, we have a hip flask containing cask strength Lagavulin 12-year-old – as things worked out, Monkey Shoulder might have been a more appropriate choice.

We start well and after passing the now boarded-up Forsinard Hotel, head back into a landscape of 90 per cent water. We find the pony track, an old, man-made path raised just high enough above the bog to make walking a bit drier. In a burst of early evening sunshine, we head off purposely in a straight line across the bogs and pools towards the Ben Griams – the prominent paired sandstone islands surrounded by a sea of peat. A double rainbow heralds our first soaking: 'Perfect peat forming weather,' I tell Bob, and he replies with that knowing smile of a brother. The Cross Lochs, famed by anglers for the size and numbers of their brown trout, appear on the horizon two kilometres away. Sir Archibald Geikie, the geologist, who came this way in 1875, was told by a local fisherman that a loch near Ben Griam was 'three parts of fish and one water'.[28]

In the post-glacial world, peat started forming here in the Flow Country on newly exposed bedrock and fresh deposits of sand, gravel and rock dumped by the melting ice sheets. It has now reached a depth of up to 3.5m and contains within it a record of change, much of it induced by humans.[29] The common occurrence of charcoal close to the base of the peat suggests that vegetation was burnt, probably by humans, as early as 7,500 years BP. With the trees gone, the ground became wetter and the growth of peat accelerated. Scientists studying the stratigraphy of the peat, layer upon layer, believe that the Flow Country has been largely treeless for the last 4,000 years, and the characteristic and beautiful bog pools systems unchanged. Annie Proulx, in *Fen, Bog & Swamp*, called these 'sinkholes into the underworld'.[30] The natural equilibrium came to an abrupt end in the 1980s with large-scale drainage of the peatlands to pave the way for a new wave of man-made commercial forests. Non-native species, sourced from the northwest coast of America, were planted here because they grew fast in the wet Scottish climate. Many of these forests were never allowed to reach maturity and have been cut down and ripped up in an effort to put right a wrong. Trees struggle to stay upright in a soil containing 90 per cent water, and when they were tall enough, were regularly flattened by the high winds that are a feature of these flat, wet lands. Forty years after they were planted, the white, bleached bones of a dead commercial enterprise litter the surface of the peatland landscape.

This is a part of the Flow Country I know well as a scientist. In the middle of a bog that escaped the forestry ploughs and held onto its pools of brown peaty water is a small aluminium construction, a flux tower, with a seemingly random array of instruments attached. Huddled close by are five solar panels pointing to the sky. A miniature wind turbine spins in the breeze capturing more energy for the instruments. Every thirty minutes, the flux tower sends a single pulse of data via satellite to a research laboratory in Edinburgh. The data pulse, which contains measurements such as wind speed and solar radiation, also includes the carbon dioxide concentration measured close to the bog surface. Over hours, days, months, years and decades, scientists are recording how the concentration of this greenhouse gas changes, and the rate at which carbon from the atmosphere is fixed by the bog. Essentially, they are measuring the growth of a peatland.

The twenty-four-hour cycle of a peat bog is much like that of a single human breath. It takes us about four seconds to inhale oxygen and exhale or respire carbon dioxide as a waste product. A single bog breath takes much longer. During daylight hours, bog plants draw down carbon dioxide from the atmosphere and fix it in their tissues by the process of photosynthesis. At night and in the absence of solar radiation, photosynthesis switches off and respiration takes over, releasing carbon dioxide back to the atmosphere. Plants also respire carbon dioxide during daylight hours, but the amounts are small compared to photosynthetic uptake. This lonely flux tower in the middle of the Flow Country, much loved by low-flying RAF aircraft, is recording the change in concentration above the ground surface as the peatland 'inhales' carbon dioxide during the day and 'exhales' it at night – a single twenty-four-hour bog breath.

PEAT AND WHISKY

As we pass by the flux tower today in early spring, we are close to one of the two most significant moments in the seasonal carbon cycle of a peatland; the other occurs in autumn. Quite soon, the bog will become bright and breezy, alive and noisy with new and returning life, a wonderful time in the peatland year. This evening, the bog plants are largely dormant, in late-winter mode, and the bog is still losing carbon each day. The peatland is coming to the end of its silent season, with the processes of carbon capture and storage yet to kick in. Very soon, as the bog is warmed by longer days and the power of the sun, the plants begin to grow again and the bog becomes greener and more productive. The silent season is over and the moment has occurred when, each day until the autumn, peat will grow again. A switch has been turned on when the bog begins to store more carbon than it gives up. In three months' time, it will be midsummer in the Flow Country, when the sun only dips below the horizon for just a few hours. This coincides with the moment in the annual carbon cycle when the bog draws down more carbon dioxide from the atmosphere that at any other time of year.

Autumn is a magical time to be out and about on a peat bog. As plant growth begins to slow, the surface of the bog undergoes a transformation in colour, the bright chlorophyll greens of the growing season changing to the yellows, oranges and reds of the carotenoids and flavonoids. Triggered by the dimming of the sun, the formation of these complex organic compounds in the plants' tissues signifies the onset of winter. It also coincides with the moment when carbon capture and storage is switched off again for the year. The bog now enters a period of relative dormancy: the long winter silent season has begun. In the short days of winter, when the sun only rises above the horizon for six hours, this is a very special and cold place to be. The water table is high and the frozen bog surface crunches under your feet; the peaty pools are full and have developed complex layers and beautiful patterns of frosty ice. Everything turns to shades of brown. The birds have gone too; the only noises are the scrunching of boots on ice and the wind gusting across the frozen vegetation. Small groups of deer move and stop, move and stop, disturbed by the distant lone intruder, as they seek nourishment in the brown winter peat-scape.

At the end of the year, the daily losses and gains of carbon measured by the flux tower are tallied up by the scientists to produce an annual carbon balance. It is an exercise in carbon accounting that reveals whether the bog is fixing or leaking carbon to the atmosphere. A healthy peatland that each year gains more carbon than it loses has a positive carbon balance and is steadily becoming deeper. An unhealthy peatland will release more carbon than it gains and have a negative annual carbon balance. If these annual losses continue, the bog will degrade, becoming incapable of holding on to the carbon it has captured from the atmosphere.

Flux tower scientists have found that this unforested, near pristine peatland in the middle of the Flow Country has a good positive carbon balance.[31] It is a healthy bog, fixing large amounts of carbon dioxide and growing in thickness at a rate of 2.5mm per year, significantly more than many other sites in the UK and Ireland where equivalent measurements are being made. This is a good news story and shows how effective a healthy, or indeed a restored peatland, can be at fixing carbon.

We leave the flux tower behind and follow a sluggish, pungent stream called the Allt a' Bhreun Bhad, the 'Stream of the Stinking Putrid Turf', before making camp in fading light in a forest plantation on glacial sand and gravel. During an unseasonably cold night, nature calls, and outside the tent in the pitch blackness, the clouds suddenly clear to reveal an infinite number of stars with the Milky Way and Orion clearly visible amongst the white noise of stellar light. Later in the night, waves of sound start to transmit through the forest as the trees begin to move. The wind is picking up and the tent becomes a target for heavy raindrops and bursts of hail buckshot. An interesting day lies ahead.

After a restless and uncomfortable night, we set off again at 8am amongst gathering showers of rain and sleet. In daylight, the forest reveals itself to be a pleasant twenty-year-old plantation of Lodgepole pine and Sitka spruce. These younger man-made forests were far gentler on the environment and easier on the eye than the blanket monocultures planted in the 1970s and 80s. New methods of contour ploughing had been introduced to hold water longer on the slopes, and tree-free buffer strips were established to protect the watercourses. Areas of birch, willow and rowan were mixed in to break up the evergreen blanket and create a more natural mosaic of trees.

The forest road cuts deep through the blanket of peat and we fill our water bottles for the day from a dark-brown peaty stream. It is claimed by some that the taste of peat in whisky is in part due to the use of peaty water during its making. I have drunk and tasted peaty water from many bogs all over the world and I have failed to taste much at all. Sometimes, there may be a slight dryness in the mouth, like tannin-rich black tea, and at other times the sepia brown water might feel smooth and soft, but never hard. This morning is no different. Flow Country peaty water, like its peat, is tasteless.

We break out of the forest onto windy open moorland and begin to ascend Ben Griam Beg, which periodically disappears in squalls of sleet and snow and moments later reappears in bright sunshine. The wind is blowing in cold, clear air from the north-west and today we are going to get assaulted by the weather. The Beg is the smaller of the two Ben Griams, geologically younger Devonian sandstone outliers or rock islands that rest on top of the older Moine Schists. Geikie visited them on a Sunday in 1875, 'a day of sunshine, of white floating clouds, and of blue distances stretching away from the purple moors to the sea'. Afraid of Sabbath breaking, his innkeeper could only be persuaded to drive him to the foot of the hills while the locals where at prayer – this was Sabbatarian country after all. It was worth the effort and he enjoyed 'a perfect weather day' on the Griams. On the summits, he would have been impressed by the fine outcrops of sandstone and pebbly conglomerate, rocks that at one time would have covered all the land stretching from here to the North Sea. The Griams are now all that is left of this great thickness of sedimentary rock that was removed by the ice to expose the older metamorphic rocks beneath.

At 580m, the rocky summit of Ben Griam Beg is crowned by the highest and most remote early Iron Age hill fort in Scotland. Piles of sandstone slabs arranged in curved lines are all that remain of what would have once been much higher defensive walls. More than 3,000 years ago, people and their animals would have periodically taken refuge here and looked out as we do now, across the treeless peatland landscape of the Flows for their would-be attackers. This was no place for a permanent residence. On the day we visit, a Sabbath day and almost 150 years after Geikie, we also take refuge from the wind, sleet and snow whipping across the summit. We are not alone and are joined by two resident ptarmigan in their transitional half-winter, half-summer plumage: a random feathery confusion of white, grey and many shades of brown. Unlike us, they seem completely at home in this high place.

In between showers, we can see the flux tower that we passed yesterday evening and when the clouds part further, the full size of the Flow Country becomes apparent. Twenty-five kilometres to the southwest, the distant peaks of Morven and its neighbours are apparent across a vast, uninterrupted, open land of flat peat bogs, pool systems and lochs. In terms of extent, there is no comparison in the UK: 'Probably in all Britain there are no scenes more impressive in their barren desolation than the enormous "flowes" [level bogs], which lie between Strath Naver and Strath Halladale, and thence stretch eastwards almost to the coast of Caithness.'[32] Very soon, big squalls of wind – sometimes carrying hail, sleet and snow, other times carrying no more than a threat – resume their assault on the hill fort and the red sandstone begins to turn white. It is time for the Sabbath-breakers to leave. We walk down wet and tussocky slopes towards the sanctuary of a boathouse on the shore of a large loch, in between showers visually locking on to the distant target of its bright orange roof. Before we get there, we are blown sideways by a westerly gale as we trudge around the fine sandy beaches of the loch. Tree roots and trunks of ancient pines, one clearly hand-cut by an axe, emerge from banks of deep, eroded peat.

The boathouse smells of fresh paint. In a calm interlude before the next snow shower, we draw breath, eat chocolate as you do at such times, weigh up our options and decide on a track that eventually takes us to a road that skirts the edge of the Rimsdale Forest. A sign attached to a deer fence suggests it is still managed by Fountain Forestry, one of the private forestry companies that were at the centre of the afforestation of the Flow Country peatlands. In complete contrast to the younger forest where we had camped last night, this was tree planting at an industrial scale. Behind the tall fence, parts of the forest have been cut down and not replanted; other areas are completely unmanaged. Defoliated and dead trees show the scars of a serious outbreak of Pine Beauty moth larvae that decimated plantations in the 1970s. It is not a pretty sight.

There are few places to camp in this open peatland and even fewer natural barriers to escape the wind. Sleeping on a peaty waterbed might sound attractive, but it is ill-advised as peat is not good at holding onto tent pegs in a wind. We are fortunate to find Rimsdale House and optimistically hope to escape the wind as best we can in the shelter of its walls. This was a small home built

for a shepherd to watch over sheep after the nearby township of
Rimsdale with its families, arable fields, enclosures and houses
was cleared in the year 1814. It was a time of great hardship in
Scotland as a period of wet summers had made peat cutting and
drying difficult, so fuel was in short supply. The shepherd and his
sheep have all gone, the house is run down and broken, rusting
pipes hang loose and the windows are boarded up. Broken glass
litters the ground and what is left of the house smells of birds
and their offspring. Red deer are now the only occupants of the
Rimsdale clearance village.

Two miles to the north of what remains of the footprint of
Rimsdale, two placenames jump out at me from the map: Allt
Bothan Uisge-beatha and Cnoc Bothan Uisge-beatha, the stream
and the crag of the whisky bothy. Ordnance Survey maps have carried these names since the 1880s and I wanted to explore this close
link between a clearance village and the making of illicit whisky.
All the ingredients to make whisky were here – people, arable land
to grow barley, high-quality peat, abundant fresh water. Distant
from Scotland's cities and towns, this would have been a small,
discreet operation – large enough to keep the villagers happy and
most likely the local laird.

That evening, we were too knackered to walk that extra three
miles and camp at the Bothan Uisge-beatha, but a few years later
I returned to Rimsdale, still keen to explore this place. Fortified
by a substantial fish supper from Helmsdale, a coastal town once
'noted for herring, orthodoxy and whisky',[33] I set off with my tent
for a night in the bog. It was midsummer, the peatlands thankfully
dry – worryingly dry, a local told me – and as I followed meandering deer paths across an open moorland dotted by stranded glacial
erratics, I was accompanied along the way by the melancholic calls
of the golden glover that return each year to nest in these wild
lands. Summer in the Flows is redolent of light winds, long days,

flowers, insects and smells – my heavy boots released volatile oils, crushing underfoot the leaves of bog myrtle. It has a heady, sweet, lemony aroma and a natural deterrent against *Culicoides impunctatus*, the Scottish midge.

I was concerned that the whisky bothy might be difficult to find, but I need not have worried. All paths leading north from the old township seemed drawn towards a point on a small rise on the horizon. When I reached it and looked down into a small steep-sided valley, there it was, the neatly concealed ruins of the Bothan Uisge-beatha and the burn that bears its name. One hundred metres above were the rocky outcrops of the Cnoc Bothan Uisge-beatha. I was delighted to be here and after evicting the local deer population, settled in to explore and make camp for the night. No whisky would have been made here during this time of year – it was too warm, water levels too low and there were more pressing tasks for the villagers. The number of abandoned shielings suggest that shepherds once came here with their sheep. Now it has become a resting place for deer.

The most noticeable ruins are two small buildings that straddle the peaty stream. With a keen eye, you can make out an excavated channel that would have taken water from higher up the burn to one of these and also carried waste away downstream – a man-made lade. Close by is a deep sunken pool surrounded on three sides by well-made stone walls – the stream had clearly been widened, deepened and the natural pool protected – probably a place to steep sacks of barley and collect water. The occupants were hydro-engineers as well as distillers.

Later that evening, accompanied by a pair of binoculars and a hipflask, I scrambled up the ice-scratched whaleback that is the Cnoc Bothan Uisge-beatha to get a better view of my neighbourhood. This was certainly a fine rocky lookout – the view is immense and even in this protracted period without rain, I can see in the silvery evening light the streams that link together a disordered scattering of lochs, lochans and pool complexes. Some of the elements needed to make whisky are still here in the landscape – water, peat, roads, but others are gone – the barley of the home fields, the crofters and their homesteads. I can't see any romance in making whisky in this tough landscape, but as the midsummer sun tracks round from the west across the slopes of Ben Loyal towards the northern horizon and begins to sink down to the sea, it is easy to think otherwise. As the wind drops, the bark of a distant deer and the chuckle of a red grouse heighten the sense of space and loneliness, sending me back to my tent amongst the ruins of the whisky bothy. The midges are swarming and I am scratching, but it is good to come back and rest a night in this place. Back in the cold spring of 2019, I had written about a quite different nocturnal experience down in the 'shelter' of Rimsdale House:

During the night, our tent receives a battering of almost biblical proportions as gusts of wind funnel round the walls of the old shepherd's cottage, periodically creating a vortex of sound and energy. Tent pegs get ripped from the ground and we secure the guy ropes with any large stones we can find in the dark. Head torches flicker on and off as we watch the aluminium alloy poles flex alarmingly under the strain before a brief lull heralds the next assault of the wind. In the morning, we wake to the sound of silence and still air. In the distance, glowing red in the early morning sun, is a fresh covering of snow on the summit ridge of Ben Loyal. A v-shaped flight of whooper swans rise up from Loch Rimsdale, whooping their way in formation slowly over our battered tent as we eat porridge and drink a second mug of tea. I count thirty of them as they head north and land close to the Allt Bothan Uisge-beatha. The paramedic tends his blisters, we pack up and head west.

There are no signposts to Truderscraig, nor does it appear on any tourist trails, maps or guides. It hasn't been excavated for our benefit and there are no helpful information boards or leaflets to tell its story. For the last forty years, it has been cut off from view, encircled and entrapped by a recently planted conifer forest. That morning, it took us three hours to walk there, firstly along a well-made track built long ago above wet bog, and then uphill through a forest of tall trees. On our way, we were accompanied by the alarm calls of chaffinches and great tits, who noisily signalled our progress to their neighbours and kin as we pushed deeper and deeper into the forest.

Finally, we break out into a large open space of grass and low walls and arrive at Truderscraig. Probably derived from the Norse *trud* meaning 'gardener', and *craig* meaning 'crag', the different groups of settlers who lived here from the Iron Age until the Highland clearances had chosen their home well. The sun came out and for the first time since we left the sanctuary of the train, we felt warm. The garden soils were good, the village faced south and west into the afternoon and evening light, and there were thick deposits of peat in the valley below. Truderscraig was one of the many Strath Naver eviction villages, where crofters were forcibly removed in 1814 to make way for large-scale sheep farming, the same year that the smaller township of Rimsdale, where we had camped the previous night, was cleared. Newly demarcated clearance lots were created on the coast to receive the inland refugees. People who previously had reared livestock, primarily the black Highland cattle, were to become coastal farmers growing crops and catching fish. The menfolk, who would leave the sanctuary of Truderscraig in the spring and summer and travel south for seasonal employment to supplement the family income, had to find new work in a new world.

Stored at Dunrobin Castle and prepared for the Duke of Sutherland's estates on the eve of the evictions is a detailed report, survey and mapping of the Strath Naver villages earmarked for clearance. The report states that the farm of Truderskeg, as it was called then, had 3,042 acres of mostly hill grazing and fifty-four acres of arable soil.[34] It would have been classed as good sheep grazing by the surveyors, who recorded that the village and its land supported and sustained eighteen crofting families in 1806. The document, which was clearly used to plan the removal of the villagers to facilitate the arrival of sheep, reveals in stark detail the calculated and systematic way that this was done, but not the human cost and the emptiness that followed. It also shows the scale of the operation.

Around the time Truderscraig was cleared, some twenty miles distant, John MacCulloch, a geologist tasked with producing a map of Scotland, was descending slowly towards Loch Shin.[35] He began to come across 'symptoms of human existence', a rough path, a ditch, the marks of a plough and even a stray horse. Arriving at a bend in the path, he saw the village. 'A shapeless heap of black ruins. All was silent and dead,' he wrote. 'These are the former hamlets of the idle and useless population of the hills: the people have been moved; but that affords little consolation to him who is thus left to struggle through these empty wastes.' A self-centred view, but one that shows that the Highland Clearances meant different things to different people. Artists like William Turner, in his 1808 mezzotint and etching *Peat Bog, Scotland*, also helped to portray this picture of poverty, hardship and gloom amongst the Highland communities. Joseph Mitchell, the engineer, had a different view, referring to the displaced Highlanders as 'brave and industrious' and not 'ignorant and idle'. In *A Skye Eviction*, Sir Archibald Geikie, in some of his most heartfelt writing, described after a day at the rocks returning to hear 'strange wailing sounds reaching my ears', and then seeing from a vantage point 'a motley procession winding along the road' of three generations of crofters, 'old men and women, too feeble to walk, who were placed in carts'. They were bound for Canada. Returning later, he found 'not a soul is to be seen there now, but the greener patches of field and the crumbling walls mark where an active and happy community once lived'.[28]

The remains of the village of Truderscraig were largely saved from the forestry plough. Still visible are the collapsed walls with gaps once occupied by swinging wooden gates, the moss-covered shapes of long houses with a smaller second room at one end for the cattle, entrance doorways and a network of paths connecting homes and extended families. Truderscraig had a burial ground, a kiln for drying grain, livestock enclosures, and on the best soil at the foot of the slope, arable land with plough rigs and furrows to grow their crops. People were forced off this land 200 years ago, but the dockens and thistles that grow and flower each year are a clear sign of past cultivation. Fresh molehills reveal a dark, fertile soil with a textbook crumb structure made by earthworms; a rich garden soil that can only be created by humans. In my mind, the evidence on the ground at Truderscraig bears little resemblance to this description of life in the townships:

> ...a half-naked hungry people living in mud-floored hovels, peat smoke blackening their faces, before it drifted out through a hole in the dripping roof, the sole room shared all winter through with starving cattle.[36]

So, what does the future holds for Truderscraig? Piles of newly arrived stock fencing suggest this fertile former garden may have new residents soon. We come across a large wooden table covered with an assorted collection of the severed heads of red deer, lower legs and entrails – a macabre and pungent offering to the gods. We leave the village, make a way through the enclosing forest and walk out into an open valley past the old peat banks where the villagers would have come in late spring to cut turfs for fuel. There are groups of *dubh* lochans, black peaty pools; some contain a little water, others are dry. They are becoming less common now as we venture further west and the landscape begins to change. Fast-moving glaciers begin to reveal their work, in stark contrast to the more static movement of the ice sheets that flattened and planed off much of the Flow Country. This is the first steep-sided glacial valley we have seen on our journey. Earlier in the day, close to Truderscraig, we came across two whalebacks, bare rock outcrops polished and smoothed by moving ice.

We have now reached Loch Choire, one of the Duke of Sutherland's former estates, and along the well-made track, feeding stations and their pungent smells herald our arrival in red deer country. A small bright green area of improved grassland with a single flock of sheep suggests a nod to the past. Visitors to the old hunting lodge at Loch Choire included King Alonso of Spain, who came to kill stags, and Winston Churchill, who came to sketch. Peat banks at the head of the loch are still being actively cut for fuel. Tonight, we find shelter from the winds blowing down the loch in a sheepfold just above the cottage of Alltalaird. In 1875, it was home to a shepherd, Mr Robert Waugh – now it is occupied by a seasonal deer stalker.

Between us and Altnaharra at the western end of Loch Naver lies a formidable barrier to progress: Ben Klibreck, a long-ridged sandstone mountain. Today, we decide to give the summit a miss – it is concealed in fast-moving cloud and a fresh covering of overnight snow. We follow a deer stalker's path, badly eroded by tyre marks up to the 500m contour. Above us, the snow still lies and the clouds begin. Way out east from where we have come, the twin peaks of the Ben Griams are clearly visible across the open landscape. They have been a constant in our travels so far, never far from view; hills that seem to grow in size the further we leave them behind. We crest a windy pass called The Whip and begin to descend towards Strath Naver beneath a dramatic cloudscape of lenticular clouds. On the way up the mountain, a monument stands out on the ridge to our west. It marks the place where a De Havilland Sea Vampire crashed into the side of Ben Klibreck in March 1955 after the pilot had lost control of his aircraft. Both crew members were killed. It is one of many fatal post-war and wartime aircraft crash sites that litter the mountains and peatlands of Scotland.

We now are entering an area of the northern Highlands that was once famed for its illicit whisky makers. As the years following the 1823 Excise Act saw increased activity on the ground by the excisemen to suppress illegal activities, the smugglers had to become smarter and more vigilant in their efforts to avoid detection. In 1888, information given to the excisemen stationed at Brora on the east coast triggered a large-scale search of the Duke of Sutherland's estates around Ben Klibreck for the smugglers and their stills.[37] They searched the lands as far as Altnaharra but found nothing. It was a hoax, a common method used by the illicit whisky makers to distract the revenue and dishearten the gaugers. Although legal production of whisky was on the rise at the end of the nineteenth century, there were still hotspots of peat-reek production in the remoter parts of the Highlands, and this was one of them. Down by the lochside, we walk through Klibreck Farm and its fertilised, unnaturally green grass – no barley, but fields full of sheep down from the hills for the spring lambing season. The farm, with its 1,000 head of sheep, is the first sizeable one we have encountered on our travels.

Our destination tonight is the Altnaharra Hotel. Less than 100m from our goal, we meet a local man and his two friendly black dogs, the first contact we have had with humans or canines for three days.

'Where are you heading, boys?'

'To the hotel, over there,' we reply.

'Not far to go, then.'

The Altnaharra Hotel is everything an establishment built for fishermen and stalkers should be. A big hearty log fire, excellent food and drink, walls bedecked with large fish in glass cases frozen in time with other assorted pond life, framed collections of flies, wall-mounted rods and stags' heads. It is still early in the season and we share the hotel tonight with three fellow guests who have been coming here to fish the local lochs and rivers for fifty years. They are polite, friendly and extremely well-dressed and spoken, but become troubled when we reveal our journey, mode of transport and sleeping arrangements. 'Henry, do you think tents are better made now than they used to be?' After three days of freeze-dried and powdered food, we drink beer from Orkney, wine from Argentina, whisky from Scotland, and eat scallops, venison and a selection of local cheeses. We sleep well. Joseph Mitchell, one of Telford's engineers, also rested here in 1838, dining on 'fresh loch trout, deer's tails, deer's tongue and venison collops'.[38] His description of an area dedicated to sporting activities suggested little has changed in almost 200 years.

* * *

Altnaharra is close to the dead centre of the northern Highlands of Scotland, twenty-five miles away to the north, east or west, lies the sea. Apart from the hotel, it has a scattering of occupied houses, an estate office and a recently closed primary school adorned with a colourful child's painting of what we both agree is a reindeer. The other major structure of note is the church, closed due to a collapsed roof. On the road out of town is a high green barbed wire fence enclosing a tower, an array of instruments, some of which I recognise including at its centre a white rain gauge. This is the famous Altnaharra weather station run by the Meteorological Office from Exeter. Famous because Altnaharra often features in the daily weather extremes for the United Kingdom. On 30th December 1995, the UK's lowest ever recorded temperature was measured here, -27.2°C, equalling the previous record set at Braemar. On 20th March 2009, Altnaharra recorded a temperature of +18.5°C, the warmest place that day in the UK. This morning, the air temperature is +8°C and rain is on its way.

The name Altnaharra is derived from the Gaelic meaning 'burn of the shieling' and at one time was a place of intersecting seasonal drove roads that were used by crofters to drive their livestock to market. It lies at the head of Strath Naver, a long river valley, and an area that has long been synonymous with the worst of the Sutherland Highland Clearances. Along the broad strath between 1806–20, fifty small townships were depopulated by the duke and his men, along with illicit whisky makers. Fifty. The dramatic decline in the local population following the clearances led to a fall in the use of the drove roads and in the 1820s, with the people conveniently cleared away, the infrastructure began to be improved. A new hotel was built to attract the upper classes to the area to fish and shoot. Thomas Telford was commissioned to build a bridge to carry the new parliamentary road,[39] linking Bonar Bridge in the south to Tongue on the north coast. These improvements were not coincidental, and not just taking place in Strath Naver. It was part of larger plan to reorganise, redevelop and support the occupation of the Highlands and Islands of Scotland by the wealthy few, their estate workers and shepherds. Lying just 43km (27 miles) to the southeast of Altnaharra, the building of Clynelish Distillery in 1819 by the Sutherland Estate was also no coincidence.

We leave Altnaharra and head west along a road that steadily climbs up and travels across a barren, featureless landscape of moorland and fenced blocks of forestry. The wind strengthens and the rain starts to set in. The weather matches the bleakness of a landscape rooted in its underlying geology – the Moine Schists – acidic, uniform, poorly drained and seemingly never-ending. We walk on, keen to arrive at somewhere more interesting. In the gloom, we can just about make out a small area of *dubh* lochans; the last remaining traces of the Flow Country. We disturb a small group of sheltering whooper swans on Loch Meadie and they skate off along the water surface, unwilling to get airborne in the strong, gusting wind. A spectacular technicolour rainbow arches itself across the loch from one shore to the other and follows us as we walk across a high plateau of glacial debris. We begin to descend towards Strath More and in the distance, out of the rain, appears Allnabad, an isolated and roofless stone cottage beside the road. Like the local red deer population, we seek and find temporary shelter here. At the time of the 1881 census, it was occupied by a family from Tongue. A shepherd, John Mackay, aged thirty, lived here with his wife Johan, aged twenty-six, son William, aged four, and daughter Annie, aged two. Ten years later, the family was gone.

We are now accompanied by the constant noise of water on the move and have entered a different land of fast-flowing rivers and streams, deep ravines and rocky outcrops. Gone are the open peatlands of the east and the change in landscape lifts our spirits. It stops raining and we both feel energised by the white water rushing past us from the high mountains of the west on its way to the sea. We are now on the old drove road to Archfary and walk along a well-maintained track to Gobersnuisgach, a hunting lodge built in 1845 with its neat, well-maintained collection of estate buildings, a game larder and a steading. We are warmly greeted by the gamekeeper and his wife, and are invited to camp on the bright grassy lawns of the lodge, which have just been cropped and vacated by the resident red deer population. We decline and thank them for the offer, preferring the sanctuary of Glen Golly and the certain shelter from the strong winds blowing out of the west. The writer and wanderer Tom Weir came this way during World War II whilst on embarkation leave before being posted abroad.[40] Three days after climbing Ben Hope and Ben Loyal, he wrote, 'I struck south-west for Gobernuisgach, steering by compass over the peaty trackless country deep in mist. I remember the relief when I hit the track and saw the meeting of the glens where the lodge is sited.' Glen Golly is a perfect place to camp, a steep-sided gorge fed by waterfalls that cascade down its sides into the river below. It is filled with trees, birch and holly, ferns, mosses and in these early days of spring, isolated clumps of milky yellow primrose and blue violet flower in its sanctuary.

By morning, the rain of the previous day had drained from the land and the river level had dropped. The cascading waterfalls of the previous evening were now little more than trickles. We fill our water bottles from the Golly. The rich, golden brown peaty water of the Flow Country is long gone and in its place is a much lighter water, not colourless, but still with a hint of peat. The wind will blow out of the west all day in our faces, but there will be no rain today. A dry buffeting is what we can expect as we climb up and cross a high mountain pass and begin to close in on our destination and the end of the journey. The path is well made and maintained, and we meet the friendly gamekeeper again, out and about dismantling old rusting deer fences. We try and show no emotion when he tells us that Lairg, twenty-five miles to the south, is basking in hot, sunny weather. A large high-pressure weather system was pushing the bad weather around the northern tip of Scotland, and we were trapped in the middle of the vortex.

In the clear air of the Bealach na Feithe, we crest the Pass of the Bog. From our vantage point at 450m, the summit of Ben Klibreck appears for the first time to us and further to the east, the unmistakable twins peaks of the Ben Griams. Two watchful, rounded islands surrounded by a sea of peat. Today, we encounter a type of peat not seen before on our journey. Forming around the 300m contour and above, hill peats cover the gentle inclines and even manage to hold on to bare rock on some of the steeper slopes. Peat forms here because the high rainfall keeps the sloping ground continuously wet. Temperatures are lower at these higher altitudes, slowing down decomposition and creating a set of conditions that allows peat to form. The covering of peat is gullied and scoured by erosion that exposes long black scars that look like superficial flesh wounds revealing the bare schist below. Hill peats are particularly fragile and transient soils, quite different from the peats that form in the flatlands to the east. They are also much younger, often devoid of tree roots, and began forming more recently after the ice finally retreated from these higher mountain areas. Thin, lying close to the bedrock, rain-fed, with a thin covering of heather, moss, coarse grasses and sedges, they are different and a product of this cold, wet, mountainous place.

From the *bealach*, we look down on to a wide, u-shaped valley bearing all the hallmarks of the recent past, and further on we can see the terraced moraines and gravel mounds created by debris transported by once-active glaciers. Higher up are the corries and *arêtes* where glaciers once actively crushed and transformed solid rock into boulders, gravel and rock flour. We are now amongst the mountains of north-western Scotland that capture the moisture of the Atlantic air and turn it into fast-flowing streams that quickly become rivers. We are crossing one of Scotland's major watershed divides, the point where over just a few steps, water that was flowing downhill to the east and then north towards the coast suddenly starts to flow west towards the Atlantic Ocean.

At the pass, we get one last view of the Ben Griam peaks; two beacons of the Flow Country. Without the aid of map or compass, they would have been like two navigation buoys to our predecessors; a ship in rough seas, disappearing from view in valley bottoms, reappearing again from a vantage point at the crest of a hill or the top of a mountain pass. As we drop down towards Loch Stack, we finally lost contact with our two mountain beacons and the peatlands of the Flow Country.

The ascent towards our evening camp is so rapid that in an instant, we cross one of the most important geological boundaries in Scotland. The Moine Schists that had provided the bedrock for our walk across the peatlands for the last five days have now disappeared. In their place is a group of rocks that have been so transformed as to make them almost unrecognisable. These are the mylonites of the Moine thrust fault, rocks that have been melted by frictional heat generated along a fault plain that once carried a huge slice of the earth's crust westwards. Underneath is a collection of younger sedimentary rocks including sandstones and limestones that look remarkably fresh and untouched by the geological upheaval that went on above. And a little further along our path, the first appearance of the Lewisian Gneiss, the oldest rocks in Europe, formed during the deepest of 'deep geological time', and the final piece in our geological puzzle.

We make camp for one last night on the sheltered banks of the Lone River in full view of Arkle and its waterfall that in normal weather conditions cascades peacefully down its grey, western face. This evening in high winds, it is blown backwards up the mountain and across the summit plateau, a gravity-defying white plume of fine mist. Tom Weir wandered this way after leaving Glen Golly: 'I was thoroughly soaked as I bounded down the pass towards a house marked "Lone" on the map.' He observed, 'Ben Stack, shadow rather than substance, mighty as the Matterhorn.'[40]

Tonight, Ben Stack was largely clear except for a small cap cloud – impressive, but no mighty Matterhorn.

In the morning, with the final drops of Lagavulin gone, we pack our tent for the last time and head down to the house marked 'Lone', situated on the flatland at the edge of Loch Stack. Above us are the great grey cliffs and white quartz-rich screes of Arkle. The house is in a good state of repair – now locked, it is no longer a sanctuary for tired, wet walkers and wanderers. Close to the bothy is a well-built low shed containing a pile of new peats and a byre for cattle at the rear. A little further down the track, we find a trailer full of tasty-looking tatties – the cows had breakfasted early and were not at home this morning. Next to the track in wet boggy ground by the loch is a freshly cut peat bank. Five whooper swans, welcome companions over the last few days, fly across our path heading north.

We pause briefly at Achfary to prepare for rain – the wind is rising, the cloud base lowering, and the day is on the turn. The Reay Forest Estate, owned by the Duke of Westminster, has its office and outhouses here, a neat collection of wooden and corrugated iron buildings. Achfary's once briefly famous black-and-white telephone box now no longer receives and sends calls and has become home to the local defibrillator. The duke failed in his campaign to keep the box operating – in 2007, it had been used for a total of twenty-nine calls, only three of which were paid for.[41] There is also a village hall, a post office, but the primary school, like so many others in the Highlands, closed in 2012. Later in the year on 28th December 2019, Achfary, like its neighbour Altnaharra, made the national weather news when it logged the highest December temperature ever recorded in the UK, +18.7°C.

We reach Loch More, simply 'the Great Loch' in Gaelic. A nice, unplanned piece of symmetry in our journey, as I had walked past another equally large piece of freshwater, Loch More in Caithness, at the beginning of this journey, six years ago. It begins to rain, and the driver of a local minibus takes pity on us and stops, winds down his window and asks if we want a lift. After making a risk assessment and evaluating the situation, the paramedic declines the offer and we head off on one last ascent that takes us up through the Reay Forest on a path that steadily rises into wind, rain and finally mist. We disappear into cloud and at 400m reach the highest point, Bealach nam Fiann, to find welcome shelter as others had clearly done before within the high round walls of what the map calls a shieling. It seems implausible that a shepherd would have built a shelter for his sheep in this exposed place. We are using it for exactly the same purpose it was built for – as an escape from the wind and rain for travellers heading west to Scotland's Atlantic coast.

We are now crossing the landscape of the Lewisian Gneiss and despite being kitted out in the best waterproof clothing modern technology can provide, I am literally soaked to my underwear. The paramedic is also super-saturated but has a spring in his step, as if he can smell salt in the air; he probably could if his nose wasn't full of water. West-coast rain is like dilute brine, created by the mixing together of cloud water and the moist warm water of the Atlantic – one of the saltiest oceans in the world. This not only produces salty rain on Scotland's western seaboard, but also a coastal, saline peat.

Very occasionally, the cloud base lifts and reveals the Cnoc and Lochan, or 'Crag and Lochan', landscape that characterises the Lewisian Gneiss in Sutherland and other parts of the world. These ancient Precambrian rocks produce an uneven landscape of rocky outcrops and small sunken lochs, many of which with time become vegetated and then peat-filled. A hardened scientist would have left the path at this point to spend some time examining another new type of peat, a peat forming in thousands of small wet rocky craters, miniature peaty ecosystems, but today is not a day for scientific endeavour. The path, which is slowly being washed away beneath our feet, rises and falls across a repeating sequence of loch-crag-peat-crag-peat-crag-loch, and then finally begins to descend. Still in cloud with the rain soaking into rather than running off us, we feel close to our final destination. We are now *cho fliuch ri sgarbh*, a Gaelic expression meaning 'soaked to the skin' or, literally, 'as wet as a cormorant'. Either way we were *drookit*, soaked and drowning.

Finally, steel-grey water slowly separates itself from grey cloud and one island and then another start to appear. At the end of Loch Glendhu, the Kylesku Bridge drops out of the gloom and then promptly disappears. The cloud base must be no more than 150m above the surface of the sea. We pass through a last gate, see a tractor carrying a bale of hay on its front forks, cross a cattle grid and arrive on the coast road.

Walking down the road towards the bridge, the rain begins to ease. Until it opened in 1984, there was no bridge of any sort across this stretch of water and travellers rowed across – their cattle had to swim. Vehicle ferries first appeared between the two world wars and one of the old ferries, the *Maid of Kylesku*, with its flat wooden hull and cage-like metal deck for cars and animals, still lies beached and rotting in the seaweed and rocks close to the old pier. As we close in on the bridge, the smell of the sea is overpowering; fresh, briny, sweet, moist, the smell of fresh seaweed and cracked lobster claw, with an occasional off-note of pungent sulphur. I am probably hallucinating – most people say it just smells like the sea. A raft of eider ducks, maybe twenty to thirty in number – I am too tied up to count – fly quickly inland in formation under the bridge, low across the water. It feels like a flypast in honour of our arrival. The air feels warmer.

We head to the Kylesku Hotel situated by the former ferry slip-way on the south shore of Loch Glendhu. Once a coaching inn dating back to 1680, it is a welcoming, warm, dry and comfortable oasis with an almost Scandi-style feel about it. Big woodburning stoves, great seafood and, most importantly, a drying room. The paramedic's boots had somehow survived the ordeal, just, and unlike the doomed explorer Sir John Franklin, he hadn't become 'the Man Who Ate His Boots'. We celebrate the completion of our journey with more beer from Orkney and whisky from mainland Scotland. Monkey Shoulder in recognition of the damage seven days of shouldering a rucksack has done to our upper bodies, and a Clynelish 14-year-old, a reminder of the Highland people of the past and the last east-coast distillery we saw from the sanctuary of the train before heading off into the peatlands almost a week ago. We arrived at Kylesku on Friday 29th March 2019 on what was meant to be Brexit Day. The rest, as they say, is history.

It is a wonderful feeling to be here and wait the arrival of family and friends, and very soon Paddy comes energetically bound-ing into the bar wagging his tail. We have completed a journey across the peatlands of northern Scotland from the cool-dry east to the warm-wet west. The trek has taken us from the Viking-settled farmlands of the Caithness plain to the high mountains of Sutherland and the rugged Lewisian landscapes of the west; from the Norse land to the land of the Gaels. We have arrived in a place that is about as far away as it is possible to be in Scotland from established whisky making. Like most of the mainland and islands, illicit distilling did take place in the northwest of Scotland, although the poorer soils and harsh climate did not produce a reli-able supply of barley. Coupled with a low population density, poor communications, no railway and the cost of transporting whisky to distant markets further south, the economics of even a mod-erate-sized distillery would have been marginal at best. Several small distilleries have begun to fill the gap in recent years, most notably on the islands of the Outer Hebrides – Abhainn Dearg on Lewis and the Isle of Harris Distillery. But from here to the Wolfburn Distillery in Thurso is a long 175km (109-mile) coastal drive around the northern tip of Scotland. The next large distillery to the south is Ben Nevis in Fort William, an even longer road trip.

This journey across a land of 90 per cent water that started on a dry, windy day in the flat lands of Caithness and ended six years later in torrential rain on Scotland's Atlantic seaboard has revealed much about the story of peat and something about the story of whisky. It is my personal take on the past and quite possibly a glimpse into the future. Small-scale cutting of peat for fuel, large-scale harvesting for energy and whisky production; illicit stills followed by licensed distilleries. Clearance of the forests by early settlers and the massive tree-planting programmes of the late twentieth century. Crops of trees and crops of grass. The 'improvement' programmes of the post-war years. Drainage followed by peatland restoration. Disposal of 'hot' radioactive waste. The eviction of Highland villagers to the coast to make way for sheep. A clearance distillery. The arrival of Victorian ladies and gentlemen to shoot and fish. People replaced by sheep replaced by deer. Evicted farmers displaced by shepherds displaced by gamekeepers. Power from peat, and power from wind. Icelandic ash and nuclear fallout. Scientific research and forensics – the signal and the noise.

Martin Martin travelled to the west coast of Scotland by boat, Boswell and Johnson by foot and pony, Alfred Barnard and his companions journeyed to the distilleries by boat, train and horse-drawn carriage. Like all good geologists, Sir Archibald Geikie liked to walk and observe in all weathers, and so did we.

Seven

Whisky Water – Pure, Soft and Peaty

*'The troublesome characteristic of peat is its
enormous capacity to hold water.'*
Peter M Dryburgh, engineer, 1978.[1]

Whisky Water

On a late summer's day in the mid-1880s after visiting the
distilleries of Orkney, Pulteney and Gerston, Alfred Barnard
travelled south through the peatlands of the Flow Country
along the recently opened Highland Railway. Passing through
the remote stations of Altnabreac and Forsinard and on down
towards the 'German Ocean', he wrote that he found the three-
hour journey 'tedious' and 'with our faces towards Inverness and
more civilized parts, we bore it with equanimity'.[2]

He was on his way to visit the Old Clynelish Distillery in the
coastal town of Brora, even then a maker of a much sought-after
single malt whisky. As ever, he was meticulous in his notetaking.
On water and peat, he wrote they 'are cut from a moss about
a quarter of a mile away, and from this moss is also drawn the
water required for motive power'. The water used by the distillery
came from the Clyne Burn, 'a small stream, which originates in
a loch, some miles distant amongst the hills'. The stream 'runs
alternatively over moss and gravel until within about a mile of
the distillery, when it enters a rocky gorge, and, tumbling over
several falls, is caught at the foot of the hill in a substantial stone
cistern, from which it is led to the distillery in iron pipes'.

At one time the Old Clynelish Distillery – now called Brora –
had its own waterwheel that powered a rolling mill to grind malted

barley and turn the arms of the mashing machine that made the sweet, sugary liquor needed for fermentation. The distillery's production and cooling water for its wormtubs came from the Clyne Burn and, when required, from a small covered, top-up reservoir that was fed by seepage water from a nearby peat moss. Downstream of the distillery, the Clyne Burn acted as a conduit to carry away the liquid wastes that were dispersed into the nearby sea. Water issues were always a concern for the distillery and in 1961, the owners reported that the 'existing water supply to this distillery cannot be depended upon to fulfil all requirements during the production period'.[3] As in many distilleries of this time, production was seasonal and stopped in the summer months as water supply levels dropped and distillery workers returned to work their crofts on the coastal plain. Water quality suffered as cool Highland streams became warm summer trickles, clear water became coloured, and purity gradually deteriorated. When a new and larger distillery was built at Clynelish in 1967 with six shiny copper stills, a more reliable and plentiful water source was a priority. The Clynemilton Burn was brought on stream, a larger fast-flowing burn that originated in Loch an Tubairnaich in the peat-rich hills to the north. A weir was built and the new distillery connected to the burn through a long pipe network. Old Cynelish was able to tap into the new Clynemilton Burn supply when their own supply levels were low.[3] Like so many distilleries in the Scottish Highlands and Islands, Clynelish has its own story of whisky and water, and like many others, it involved peat.

Sunday was a quiet day at Clynelish Distillery and a good day for a walk into the hills in search of distant lochs and bogs. The September berry season had arrived and as I wandered along the hedgerows by the Clyne Burn past the old clearance lots, blackbirds and mistle thrushes feasted on an abundance of red rowan, black brambles and intensely orange rosehips. A solitary swallow lined up on a telephone wire, alert and patiently waiting for company. Sunshine and showers was the forecast for the day, a welcome return to normality after yesterday's deluge. After a 'true Highland breakfast', I walk lazily along the old coffin road that follows the burn and winds its way up past the ruin of Clynekirkton Parish Church and its watchtower that looks out over the flatlands and clearance lots below.

Highland streams often naturally lose some of their peaty colour before arriving at a downstream distillery – but not today. Leaving the road behind and following the Clyne Burn uphill, flowing fast, full and brown this morning after yesterday's rain, I take a big step back in geological time to the Precambrian era. From the soft, coal-bearing Jurassic rocks of the coastal plain with its green fields and noisy sheep, my path steepens suddenly as it climbs up and traverses the hard, metamorphic Moine Schists. Next to the path lies the deep, rocky gorge where Barnard many years ago saw the Clyne Burn 'tumbling over several falls', and after I walk past a graveyard of assorted rotting and partially gutted vehicles, I arrive above the home fields, draw breath, and hop over a fence onto an old, moorland path – a peat-cutters' path. It is one of two that were painstakingly built to carry cut turfs downhill from the upland mosses to the crofts and distillery below. The width of a pair of cartwheels, it is now disused and has become a rough, sodden track, and today a culvert for yesterday's rain. The path builders had painstakingly constructed a sunken track on the hard, rocky ground found below the peat and cast aside angular boulders of grey schist that now litter the surrounding banks.

For now, my path becomes steady, never steep, and purposefully snakes its way into and through a flat land of peat. At this point, the Clyne Burn begins to slow, lose focus and meander aimlessly before being absorbed into a boggy place of tall rushes and mosses. Further along the path, the burn reappears, its flow quickening again as I head steeply uphill across rocky crags. Another leap in geological time, this time 500 million years forward, as the boulders beneath my feet change from the banded white and grey schists to the iron-rich brick reds of the younger Devonian sandstone. This is one of the world's best aquifer rocks – a perfect home for deep water. Finally, over the brow of a hill one of the distillery's two water sources comes into view.

I have reached An Dubh Lochan, the Small Black Loch, at the point 'some miles distant amongst the hills' where Barnard's 'small stream' exits the loch. The distillery workers had built a rudimentary dam of well-laid and well-crafted sandstone slabs, backed up and sealed with peat cut from the surrounding hillside. The dam clearly hasn't been touched for years but is still doing the job it was intended for. The burn was also straightened and deepened below the dam to encourage water to flow freely downhill. Kneeling on the slabs of sandstone, I lean forward towards a dark, peaty limnological world and fill my metal bottle with a sizeable slug of water. It has a pleasing light-brown peaty tinge, tastes good, and will be used to dilute something stronger later in the day.

Like much of the water in the Scottish Highlands, the Clyne Burn is pure, soft and peaty, great for making whisky and diluting a dram, the water of choice for many distilleries in Scotland. In Japan, where its distilleries often tried to replicate the methods and traditions of the Scotch whisky industry, many sourced soft, neutral or mildly acidic water, some of which originated from peatlands. Finding a suitable water source was so important for the Suntory Company that they spent two years looking before finally deciding to build the Hakushu Distillery below the soft waters of Mount Kaikomagatake on the island of Honshu.

Distilleries use water for a range of production processes, including steeping barley, mashing, cooling condensate and spirit reduction. Condensers and wormtubs require large amounts of cold water to operate efficiently. Distilleries also need water to generate steam from their boilers and for heat recovery systems, just some of the essential roles that water plays in whisky production. Water is added, and water taken away.

So why soft water, and what makes it so good for making whisky? Firstly, it has a low salt content and is slightly acidic with a pH less than seven, and contains low concentrations of ions such as calcium and magnesium, as well as micro-nutrients that are needed to promote enzyme activity and yeast growth during mashing and fermentation. In the Highlands and Islands of Scotland, soft water is a direct reflection of the acidic, older weathered rocks and the organic-rich soils that make up much of the country. This natural acidity persists during the whisky-making process – at the beginning of mashing, the acidity of the malt-water mix is generally around pH 5.2, ideal for yeast growth and low enough to promote the release of important compounds during mashing.[4] At the point of bottling at 40% abv, the pH of Scotch whisky is often as low as 4.0–4.5.[4]

Several distilleries in Scotland are different and use hard, alkaline process water. They are few in number, but notably include Highland Park and Scapa from peat-rich Orkney. Both distilleries take their water from lime-rich springs – it is old, much travelled and deep sourced. The springs are often believed to possess medicinal properties, and local people from nearby Kirkwall would collect the lime-rich water and use it as a treatment for rheumatism. On the mainland, the mineral-rich Tarlogie spring is the mythical and sacred source of the 100-year-old water used by the Glenmorangie Distillery. Further south, Glenkinchie and Loch Lomond distilleries draw their water from sub-surface boreholes and springs that tap into calcium-rich bedrock.

Hard, mineral-rich water can create problems for distillers and may need to be treated and demineralised. In limestone and chalk-rich regions, fouling of equipment can occur. Much like the formation of limescale on the inside of a kettle, it is caused by the precipitation of salts when mineral-rich liquids are heated. Brewers and distillers in hard water regions can run into problems with beerstone, a calcium oxalate salt that precipitates in fermentation vats, forming scale or crust. The high lime content of Tarlogie spring water was sometimes problematic at Glenmorangie Distillery until a filtration system was installed in the 1980s. Spirit reduction using untreated, mineralised water has been known to produce a precipitate, a 'hazy floc', that looks like tiny strands of cotton wool floating in the bottle. It has even been said that whisky in a glass could turn a 'milky blue' colour when chalky or lime-rich water was added due to a reaction with copper leached from the insides of the stills.[2]

It would be wrong to say that high-quality whisky cannot be produced from hard, mineral-rich source water, something that is particularly true in the states of Kentucky and Tennessee. This is hard water country, where it is widely used in the production of bourbon whiskey. The limestone-rich soils not only grow great corn but also produce clear, calcium-rich spring water brought to the surface from deep aquifers. The distillers maintain that hard, alkaline water both enhances enzyme activity during fermentation and is free from the iron that can cause discolouration. Jack Daniel's strongly promote their hard water source and state that it is almost 'sterile and iron-free' and unsurprisingly 'makes the best whiskey'. Many American and Canadian distilleries also use limestone water as they feel it enhances the formation of key congeners important in producing the flavours that they want. In some cases, the pH of the water has to be artificially lowered. The process of sour mashing, the addition of acidic spent mash to a new mash, is one way of doing this, neutralising alkalinity and importantly inhibiting unwanted bacterial growth. It is not unlike the starter process used to make successive batches of sourdough bread.

In essence, both water quality and water quantity matter to whisky makers and whisky drinkers. Water quality is essential for any part of the process where water addition occurs – malting, mashing, reduction of new-make spirit prior to cask filling, bottling and drinking. For condensation, heat recovery, steam production and cleaning, all parts of the process where water is not in contact with the raw materials used to make the spirit itself, quality can be sacrificed for quantity.

Water – Pure and Clean

I eat a sandwich by An Dubh Lochan and ruminate about the moss below on the old peat-cutters' path. From my vantage point, I can see where the Clyne Burn loses itself in the peat bog and then reappears some distance away before heading downhill towards the Old Clynelish Distillery. The disappearance of the burn into the peat bog is important to the distillery water supply for two reasons. First, the bog is acting as a natural water storage tank providing the distillery with security of supply. Second, it operates as an effective filter bed, removing, adsorbing and retaining contaminants and particles. These include airborne pollutants such as heavy metals, nitrogen and sulphur that circulate in our atmosphere before being washed out and returning to earth, often in places far away from industrial sources. The surfaces of peat particles are home to an unusually high concentration of charged chemical exchange sites that have the capacity to adsorb and effectively lock up different types of pollutants. When the water leaves this organic filter-bed after a pause in its journey, it is cleaner than when it arrived. The peat bog is acting like a natural water treatment plant, filtering, demineralising and decontaminating.

In the days when this moss was cut for fuel, its capacity to hold and filter water would have been much reduced. The tell-tale cuts, scraps and changes brought about by humans are obvious from the abrupt transitions in vegetation, jumps in colour and texture, straight lines, dark broken peat banks, rectangles of bright green grass and soft rushes, all randomly strewn across the landscape in front of me. There appears to be nothing systematic about the way peat was cut here and little or no evidence of an organised drainage plan – factors that will help the bog to recover now that it has been left alone. The peat was cut from a series of medium-sized damp depressions, each one joined by interconnecting paths. Abandoned a long time ago, the undrained bog remained wet and is now regrowing and deepening again year by year.

The bogs that provided peats for Old Clynelish formed in a series of depressions in the hills to the north of the distillery. Like many other cut-away bogs, they once had a light railway, now long gone and reclaimed by the rapidly growing moss. On my way up this morning, I had come across rusting metal poking out of the bog – a large, riveted cylinder and a peat gauging rod, forgotten, left in the same position when it was last used by someone to push down through the soft surface until it met the resistance of the deep mineral soil. Over the years, winning peats had clearly become harder and the last attempt was the least successful, leaving an ugly large erosion scar on the side of a hill. The dry peat here was barely 50cm thick and the hillside scar may never heal, marking and wounding the land forever. Peat had been sporadically worked all the way up to the loch and beyond as scarcity and need drove the peat cutters deeper and deeper inland away from their crofts. All around An Dubh Lochan, the old peat banks look like they have been gnawed by some Holocene giant, its bite marks scarring the surrounding slopes.

I can picture the old peat cutters also having their lunch on a clear day like this, in quiet conversation, eating their pieces and taking in the big skies and panoramic views to the south. The Brora River snaking its way to the North Sea through the green fertile fields of the Clyne coastal strip. The Dornoch Firth stretching out in the distance, and far off the red banded column of one of Stevenson's lighthouses, Tarbat Ness, today standing alone like a floating mirage on open sea. The first distillery workers to cut peat up here would have observed the building of the lighthouse, each month growing taller and taller, and then one early evening may have looked out again, rubbed their eyes, blinked twice and seen the moment in 1830 when the light was first lit. On days when the visibility was this good, their gaze would have travelled further to the Moray Firth and beyond to Ben Rinnes, a distant sentinel in the heart of Speyside. Like me, they may have even wondered who came here before.

Having ruminated for far too long over lunch, I walk round An Dubh Lochan and climb higher towards the source of the Clyne Burn through ridges of peat that with each step get deeper. At 300m, I come across a solitary pine stump, the first I have seen all day, poking out from an eroded peat bank. A posse of five young stags charges across my line of sight and disappears over a hill. Eventually after some wayward route finding, I reach Loch an Tubairnaich and catch up with the stags. I am surprised by the loch's size. This mass of water has been put to good use by the distillery and acts as the header tank for the Clynemilton Burn, the water supply for the new Clynelish Distillery. The peats are deeper here – uncut, intact and very wet – too far, too remote and pathless to attract attention. From its peatland reservoir, the burn charges downhill through a steep-sided valley before losing power when it meets the coastal plain and the old clearance lots at Achrimsdale, my destination. I too charge down the hill in a futile attempt to outrun an approaching afternoon storm – I quickly pass young conifer plantations, follow fence lines, open and close gates, and eventually find a new well-made track. There has been much change in the valley recently and from time to time, a bright blue water pipe pops up, fast tracking light-brown cooling and process water to the distillery below. The burn is said to contain particles of alluvial gold, from a source still unknown. I have no time for panning today but nine miles due north from here is Kildonan, the place where Robert Nelson Gilchrist first discovered gold in the burns of Sutherland that triggered the great Scottish gold rush of 1868.[5] I didn't outrun the storm and that evening, back by a warm drying log fire in the Clynelish Farmhouse, I write up my notes of distant lochs and bogs with the help of the local dram, diluted with pure, soft peaty water from An Dubh Lochan.

Distillery owners and their enthusiastic marketing departments are rightly proud to promote their water sources, but it would be fair to say that there is only one place where more nonsense is written about water, and that is in the mineral water industry. Pure, clean, natural, crystal clear, sparkling, alpine, mountain, glacial, unique, and even wild – all words that regularly find their way into the lexicon of water speak, but surprisingly not 'peaty'. Over-hyping your water source is not just a recent phenomenon; Alfred Barnard, writing about the days of the whisky smugglers, or the 'Pioneers of the Whisky Trade' as he called them, said they chose 'places where the purest mountain streams, flowing over moss and peats, could be used to distil and produce spirits of the finest descriptions'. Neil Gunn was a great exponent of the quality of his native Highland water, saying it was the 'best water for making pot-still whisky'.[6] In a 1989 magazine advert, the marketing department of the Glenlivet Distillery had their kilted distillery stillman and resident sage waxing lyrical, standing astride a foaming Highland stream with a glass of Glenlivet in hand uttering a toast: 'Some say it's the smoke of the peat cut from the nearby Faemussach peat fields. Our own Sandy Milne insists that it's water from Josie's Well, flowing as it does down through peat and over Highland granite.'[7]

Since whisky is commonly bottled with an alcohol content in the range 40–46% abv, in most cases water is the dominant liquid component. Scotland is blessed with water of high purity, and for this reason water treatment was traditionally either not required or kept to a minimum. Water quality is so important to the whisky industry that some companies, like William Grant & Sons, go to the considerable expense of tankering Speyside water from their Robbie Dhu spring to their blending and bottling plant at Bellshill near Glasgow. At one time, water from the Grampian mountains was being brought to London in large porcelain tanks by a blending company to reduce spirit prior to bottling.[6] There were even rumours that it was finding its way overseas. This unpatriotic activity horrified passionate locals:

> Are the Scots now utterly spineless? The news that our famed waters are being exported to London and the Continent to make whisky in competition with the real world-famed Scottish product must have come as a shock to thousands of Scots... What a lot of suckers we are!
>
>   Yours faithfully, Tain o' Shanter.[8]

It's hard not to be awed by some of the latest claims about the magical properties of pure water now being used to promote whisky around the world. How about an independently bottled Port Charlotte single malt whisky from Islay containing 'drops' of 1,000-year-old iceberg water brought back by an explorer from the Sermilik Fjord in east Greenland?[9] Wow. Or an 'alpine' whiskey produced by the Glacier Distilling Company using pure glacial meltwater from the Northern Rockies? Aurora Spirit, at 69°39'N the world's northernmost distillery, based in an old Cold War NATO coastal fort in Arctic Norway, using unique, filtered, glacial water from the Lyngen Alps? Its whisky, Bivrost, a Viking word for the Northern Lights, is made from water that was first frozen into mountain glaciers more than 5,000 years ago.[10] Personally, I quite like the thought of drinking whisky that has been nurtured with 5,000-year-old water.

The Braunstein Distillery in Denmark has gone one step further and uses iceberg water imported from Greenland to make its Isfjord Whisky, 'Distilled with the world's purest water':

> In 2007, Isfjord discovered the amazing characteristics of distilling spirits with one of nature's best kept secrets: the extremely pure and soft water from icebergs, that naturally breaks from the Greenland Ice Cap and into the sea in Ilulissat – Greenland far north of the Polar Circle. Water that has been preserved as ice for more than 10,000 years, hidden from the outside world, sealed from any pollutants and almost all forgotten. Just until the very moment when the icebergs scatter into millions of pieces and the local icemen harvest a few of them and melt them into the purest natural water on earth.[11]

While water just seems to get purer, older and wilder for the marketeers, pure glacial aqua is clearly not just the preserve of the global bottled water industry. But does it make a difference? Greenland iceberg water does contain very low concentrations of dissolved solids and it may, in the company's words, be some of the purest on the surface of our planet. It comes from snow that fell to earth at a time before the Anthropocene, the moment when humans started contaminating the atmosphere with airborne particles and gases. Concentrations of CO_2, measured in tiny bubbles of air trapped deep within the Greenland icecap, were at this time around 270ppm – much lower than they are today. The scientific definition of pure water is simply H_2O and nothing else – it contains no impurities like minerals, salts, nutrients, microbes or particles. It is ionically pure and has a conductivity, a measure of the ability of water to conduct electricity, close to zero microsiemens per centimetre ($0 \, \mu S \, cm^{-1}$). Ions present in rain, seawater and produced by the weathering of rocks all make water less pure and increase its conductivity.

Unless it has been polluted or contaminated, the purest natural water in the world is therefore likely to be either rainwater or melted glacial water. Water can be purified artificially by processes such as filtration or ultraviolet oxidation and vaporised and condensed to remove salts to produce the purest water made by man, known as distilled water. Laboratories throughout the world rely on distilled water to control and replicate their experiments and processes, as do whisky makers and blenders who often use it, or demineralised water, as a neutral and consistent diluter to reduce spirit strength prior to casking or bottling.

In 2017 came the news that Swedish scientists had 'proven' that whisky tastes better when water is added.[12] This is something that master blenders, distillery managers and whisky drinkers knew already as adding water opens the flavours present in the congeners. But what type of water should you add to a dram? For very good reasons, Malvern water has been promoted as the perfect water for diluting whisky in the glass and the Romans were amongst the first to recognise its quality and purity. It is celebrated in an eighteenth-century rhyme: 'The Malvern Water, says Dr John Wall, is famed for containing just nothing at all'.[13] Dr John Wall was a famous physician who recognised the value of a natural water source that was pure, plain, unflavoured and unmineralised.

Good whisky deserves more than chlorinated tap water and several brands are now targeting the more discerning whisky drinker. Marketed as 'Wild Water for Whisky', Larkfire originates from a loch that rests upon three-billion-year-old Lewisian Gneiss located on Arnish Moor on the Isle of Lewis. Sourced from the oldest rocks in Europe, the water is extracted and then shipped to the mainland for canning as 'the purest wild water'. One autumn day, I decided to track down the source of this water and followed a pair of vehicle tracks up and across a bouncy wet peat bog covered with aromatic bog myrtle. A line of posts with the occasional manhole cover led to a concrete lochside hut, the site of an old pumping station. From here, a blue pipe disappeared into the loch towards a bright orange tethered buoy, the source of Larkfire. The loch water had a slight light-brown tinge, transparent and crystal clear. It was a peaceful place to sit on a warm autumn afternoon amongst the Lewisian Gneiss – not the wild place that I had come to expect.

That evening, I opened a can of Larkfire. Compared to the colour of tap water, it did have the pleasingly light peaty tinge that I had seen earlier in the day, reflecting its provenance. I found it quite pleasant to drink on its own, still and very soft with low concentrations of calcium and magnesium but higher amounts of sodium and chlorine, reflecting its close proximity to the sea. The company Uisge Source goes one step further, marketing three types of mineral water from wells and springs in the Highlands, Speyside and Islay to specifically accompany a dram from these three whisky-producing regions.

But why not collect pure mountain water on your next journey to the hills and add it to your whisky? I learnt this top tip from Jimmy who lived at Carron Lodge in Speyside, a great man of the hills who carried an old battered silver Swiss-made Sigg bottle with him on his regular trips to the mountain. After drinking it dry during the ascent, and before starting his descent, he filled the empty bottle with soft, pure water from a point as high on the hill as possible. Back home by the fireside, he would add a dash of this natural aqua to his favourite Black Bottle Islay-heavy blended whisky, rather than the inferior stuff that comes out of the tap or a mineralised bottled substitute. Maybe the difference is in the mind. It didn't matter – the whisky always tasted better.

All water sources are of course unique, but perhaps the final word should go to Ashok Chokalingam of Amrut Distillery in southern India. Water is again to the fore in the marketing of his whisky. On the label of a bottle of Amrut single malt is the unmistakable image of Mount Everest accompanied by the words 'Made From Selected Indian Barley Nurtured By Water Flowing From The Great Himalayas', which lie some 2,500km to the north-east. Amrut Distillery in Bangalore gets its soft process water from boreholes, which then travels the final two kilometres to the distillery by truck – not such a great story for the marketing department. Ashok was once asked about water and in his own characteristic style, he replied: 'We write this story', he points to the text on the Amrut tube in his hand, 'because people want to know where everything comes from. That everything has an impact. But I can tell you now that the idea of water source imparting flavour is total fucking bullshit!'[14]

That's cleared that up, then.

'The journey of the water of life.'

The town of Fort William is a place where it always seems to rain. It sits in the shadow of Ben Nevis, Scotland's highest mountain, that acts like a magnet not only for walkers but for moist Atlantic air that frequently shrouds its top in cloud. Between 1883 and 1904, a weather station operated on the summit plateau manned by a superintendent and his two assistants.[15] For twenty years, they made hourly meteorological readings, recording an astonishing average of 4,350mm of precipitation per year, six times the amount that falls on Edinburgh. Its summit clouds mark the start of the story of 'The Journey of the Water of Life', told eloquently on the label of a bottle of Ben Nevis single malt – a tale of mountain rain and snow, peat and granite, lochans and streams, distillery pipes and whisky:

It starts as rain or snow falling on Scotland's highest mountain – Ben Nevis. Either as rain or melting snow it percolates the thin layer of peat soil until it reaches the granite rock and unable to penetrate it runs under the surface until emerging in Coire Leish or Core na Ciste. The flows from these two mountain lochans, located well over 3,000 feet above sea level, make their way spilling over the blue and pink granite rocks of the mountain's rugged north face until they join together as Allt a Mhullin continuing on in the valley between Ben Nevis and Carn Mhor Dearg. For the last 750 feet of descent this water is piped underground into the rear of the Ben Nevis Distillery where our small, dedicated workforce work their alchemy to convert this pure clear sparkling water into the crystal clear Ben Nevis malt spirit.

In his famous book, simply named *Whisky* and published in 1930, Aeneas MacDonald clearly understood the importance of that journey and the pathways along which water must travel before it reaches the distillery.[16] Summarising the main virtues of whisky water, he pronounced that it should add no flavour, 'be pure, colourless without odour, free from micro-organisms or grosser organic matter and above all devoid of mineral contamination, dissolved or in suspension'. He even goes on to highlight the importance of geology, soils and the riverbed as 'rain-water acquire(s) a common body of local or provincial qualities' as it flows downslope. MacDonald neatly connects water quantity and quality, recognising the significant transformations that take place as it journeys towards the distillery.

Unless it arrives on bare rock or open water, as soon as rain hits the earth's surface, this mildly acidic, low ionic strength liquid begins to change. The journey has begun as it infiltrates the surface vegetation, finds its way along a network of connected soil pores and pipes past the root system of plants and down into deeper soil, peat and bedrock. Some may be quickly lost back to the atmosphere by transpiration or evaporation from a leaf surface or open water, the hydrologist's angels' share. When rain meets peat, its journey is likely to be delayed. Peat has an impressive capacity to hold onto its water and after periods of sustained rain, this giant organic sponge will be full, holding twenty times its own dry weight in water. It is completely unique – no other soil or rock on our planet has this capacity to store such colossal amounts of water. If peat has a natural terroir, so does water, arguably to an even greater extent.

Water and whisky are of course linked at birth and by name, the modern word whisky being a derivation of *uisge beatha* (Scots Gaelic) or *uisce beatha* (Irish Gaelic), meaning 'water of life' or 'lively water'. The link between whisky and water is celebrated in distillery names: Wolfburn, Coleburn, Springbank, Loch Lomond, Royal Lochnagar, Reservoir, Waterford, to name but a few. Scots Gaelic has a profusion of words for flowing water, *abhainn*, *allt* or *ault*, some of which have found their way into distillery names: Aultmore meaning 'Big Burn' and Abhainn Dearg meaning 'Red River'. Distilleries like Glenlivet, Glenlossie, Glenfiddich and Clear Creek all take their names from their local river, reflecting its importance to the whisky makers.

We can also find names that link a distillery to its peaty water: Dallas Dhu (valley of the black water, or black bog) and Mackmyra in Sweden (*myr* is Swedish for bog or mire). Linkwood Distillery receives some of its water from the Burn of Bogs, and the Black Burn is a source water for the distilleries of Glen Grant and Miltonduff. Further north, water originating from the Loch of the Peats makes an important contribution to whisky made at Glen Ord Distillery.

Peat, power and water were once inseparable and at one time, flowing water provided the kinetic energy to make whisky. Nowhere is there a stronger reminder today than in the giant waterwheels of

Ireland's distilleries that now lie idle at Bushmills, Kilbeggan and Midleton. One third of the land surface of Ireland is covered in peat mosses and this giant bog water reservoir acted like a reliable battery that never lost its potential to power and drive the mash-tuns and grindstones of the distilleries. Waterwheels powered mills that spawned breweries that became distilleries: Strathmill, Littlemill, Miltonduff, Noah's Mill, Millburn and Milltown, which in 1951 became known as Strathisla, renamed after the river that flows past its doors.

As distilleries grew in size and more reliable carbon-rich fuels became available, their waterwheels slowly ground to a halt and became redundant, as did their peat mosses. One hundred years ago, the men of Glenlossie cut their peat from the Birnie Moss in the hills high above the distillery. The same moss gives birth to embryonic streams that emerge from peat banks and begin to trickle downhill, before converging, steadily growing in size and direction, eventually becoming significant enough to earn a name. The Barden Burn heads north following the Peat Road down past Shougle and by the time it reaches the distillery, it was large enough to both power its old waterwheel and provide sufficient water to make 90,000 gallons of whisky a year.[17] The Birnie Moss no longer provides the distillery with its energy – its sole current role is as an upland reservoir of water. In an interesting twist of fate, wind turbines from the same hills now generate electricity for the grid and power for the local distilleries.

Glenlossie Distillery lies within the catchment of the River Lossie, one of Scotland's most important whisky rivers. It is large, fast-flowing and flood prone, but is not in the same league as Scotland's four prime movers of water – the Spey, Tweed, Tay and Dee that all flow east across post-glacial landscapes to the sea. Scotland's watershed divide sits way over to the west, because here lay the thickest accumulations of ice. When the climate warmed, the ice slowly disappeared from the high mountains. Following predetermined paths scoured out by ancient glaciers, the melt-waters travelled east, flattening or smoothing all barriers in their way. It is the moment when Scotland became what it is today – an asymmetric land of water, lochs and rivers. Most of Scotland's water flows east, and here lie most of its distilleries.

Distilleries need water and lots of it, so as they grew in number, they became concentrated around the most reliable and largest waterways. Nowhere is this more so than in Speyside, where more than fifty distilleries are clustered into this single watershed covering an area of over 3,000km². They source their water from its tributaries, many of which start life in the peat covered hills of the Spey valley. Some extract it directly from the main stem of this single 'mother' river that starts its journey in peatlands close to Loch Spey, just 35km (22 miles) from the west-coast town of Fort William. Spey Bay, its point of arrival on the North Sea coast, lies 172km (108 miles) downstream to the east, a clear demonstration of Scotland's hydrographic asymmetry. The Spey is not only one of Scotland's largest and longest rivers, but with an average speed of flow of 36mph (56kmh), it is its most powerful. It is not just a mecca for whisky people and fisherfolk, kayakers love it too.

On a cold winter's day in December, I set out to search for the source of the Spey. I leave the well-gritted public road at Laggan and drive west, upstream, following the route of one of General Wade's military roads. It's cold, really cold, and overnight snow lying on thin ice has made driving conditions hazardous. Today's forecast is good, and I press on past fields of sheep and Highland cattle, through frosty forests, on up past the Spey Dam, where half of the river is syphoned off into a canal and piped west to the pow-er-hungry aluminium smelter at Fort William. A disruption of the natural order, the Spey is uniquely a river that flows both east to the North Sea and west to the Atlantic Ocean. I drive on slowly past the old military barracks and over Wade's Garva Bridge, the first crossing of the Spey, and along his dead straight military road. Finally, I arrive at Melgarve where I leave the car, planning to be back later in the day. The general's road heads off to the north-west, following the route of one of Scotland great drove roads. The Corrieyairach Pass once brought livestock from the summer grazing lands of Scotland's west coast and islands to the markets of the central Lowlands, and would have been a convenient high-level route for the local whisky smugglers.

By now, the sun is up and the mountain tops have turned pink as I drop down to the Spey and follow a set of vehicle tracks along an open floodplain. The river has become more sluggish as it nears its source and bears little resemblance to the fast-flowing grandeur of its middle and lower reaches. I break away from the forest and arrive at Shesgnan Bothy where the tracks I had been following do an abrupt about-turn and head straight back from where they came. My footprints in the snow are now joined by those of deer, who keep watch on me from a safe distance on the far banks of the river. I walk past and over drumlins, small mounds of glacial rocky debris that were deposited and shaped by ice moving east.

I walk on across boggy, wet terrain, following an old track that becomes more snow-filled as it climbs slowly upslope. There are new tracks in the snow; the three plus one paw prints of a mountain hare and the busy contorted trail of a field mouse. Below me to the south is an ice-covered elongated loch, the source of the Spey. Or is it? According to those who argue about definitions, Loch Spey is one of three possible sources of the River Spey, the others being either the Shesgnan Burn to the north or Allt Coire Bhanain to the south, both of which are much further from Spey Bay than the loch itself. However, Loch Spey is the most prominent source, a focal point for a collection of tributaries that flow down from the surrounding bogs.

PEAT AND WHISKY

I now crest Scotland's great watershed divide and beneath my feet, water begins to trickle away to the west towards the ancient Caledonian pine forest of upper Glen Roy. Leaving the sanctuary of the path, I walk towards an area on the map that shows tributaries disappearing amongst a boggy flatland. I am walking on a bearing that is marked by an occasional rusting metal post and pieces of tangled wire, the remnants of a fence that once delineated the watershed divide. I follow a tributary that climbs up onto an open plateau and amongst this land of peat I find 'bog treasure'. Dozens of ancient roots of Scots pine reveal themselves through the snow – all that remains of a sizeable ancient fossilised forest that has been consumed through the ages by peat growth. The tributary I am following finally dissolves into small rivulets, like the microscopic capillaries at the ends of our blood vessels. I punch a hole in the ice with the heel of my boot and beneath is a trickle of water with barely enough momentum to carry it downslope. It has very little colour but tastes good. As I walk on, the ice-covered capillaries disappear and I find myself in the middle of a system of long sinuous, flooded pools, like those that characterise the Flow Country. Ribbons of frozen water, they line up one after the other, like ripples across the watershed.

I become distracted as I walk on across this bogland and my left leg disappears through thin ice into a cold, deep, unpleasant place. Wet and tired, I stop struggling across this landscape and pause to take in my surroundings. I am in the middle of a vast open peatland that extends along the watershed divide; to the west is Glen Roy and beyond Fort William and the Atlantic; to the east, Loch Spey and the start of the long river valley that stretches all the way to the Moray coast. As the grey winter sun, low on the horizon, moves steadily west, it picks out occasional ice-scraped rocky outcrops and abandoned glacial erratics that lie scattered all around. The water beneath my feet has become static and won't begin to move again until the peatland defrosts in spring. There is something ethereal and peaceful about being in the middle of a vast frozen, snow-covered peatland on a windless winter afternoon. There is beauty here in the silence; I love these places.

According to the old map makers, the bog where I am standing is so remote and devoid of anything of note that it bears no name and deserved nothing more than the words 'col, 1,151 feet' on the map. But this is not an empty place. I stop for a few moments on a rare rock outcrop high above Loch Spey, write a few notes, chew on a hard chocolate bar and watch a small herd of deer lolloping silently across the bog. The short, distant rattle of a solitary grouse pierces the silence. An occasional sound wave of fast-flowing water descending from a distant, high corrie. The weather is on the change, snow is forecast tonight and before my fingers become too cold, I return to the path and walk back down the upper reaches of the Spey. I had been warned that walking to the source regions of the Spey was a 'bog fest' and it was.

Every time I walk across Scotland's great watershed is different, and while the sources of many great rivers are clearly defined by a high mountain spring, the Spey rises in many secret, concealed places. Hard to find, almost without noticing, water oozes out of the bogs and starts to flow east as it sets off from its source pools. On its journey to the sea, the proto-Spey steadily grows in size and gains speed. After passing through the shallow Loch Spey, it is confronted and stopped by the formidable barrier of the Spey Dam before regrouping and heading north-east where it is joined by numerous tributaries, many of them originating in upland bogs. Some, like the Rothes Burn that flows past the Glenrothes Distillery, have the signature rich, orange-brown colour of peat water; other major tributaries like the Livet, Lour and Fiddich are more lightly coloured, reflecting more distant peaty source areas high in the Speyside hills.

PEAT AND WHISKY

The Promontory Hill

We woke in the morning to the flashing lights of a gritter, followed by the sound of distillery lorries passing slowly by as they carefully negotiated the icy road outside our window, their headlights creeping slowly across the bedroom ceiling in the darkness. At 7am, cars began moving up and down the road outside the old Brewer's House where we were staying – a shift change. Sofia and I had arrived in darkness and as the overnight rain and snow disappeared east and the stars came out, it got colder. At daybreak, we surfaced, breakfasted on tea, porridge and coffee, made hot drinks and food, dressed in winter gear, shouldered rucksacks and headed off up the road in bright morning sun towards the 63ft (19m) chimney of Benrinnes Distillery. It is not open to visitors, but this morning we were going higher and would be getting much colder.

The Brewer's House is just a stone's throw from the warehouses and as we walk past the office with Paddy, our four-legged friend, the distillery is already busy and up to speed for the day. From the water storage ponds, we climb uphill on the old peat cart track through fields and into a wood where we cross the lade that brings mountain water down the Scurran and Rowantree Burns to the distillery below. Soon, we are on open moorland and an icy track that is becoming more treacherous by the minute – we pause long enough to add mini-crampons to the soles of our boots. With bright low sun in our eyes and a cold wind from the west that increases in velocity as we head higher towards the snow-covered summit ridge, we climb further on the ice with more confidence. It feels good to be alive today.

The visibility is excellent and as we climb, the view gets bigger and better as Speyside's distilleries begin to appear one by one: Cardhu, Dailuaine, Aberlour, Macallan and Craigellachie. Beneath us, the valley of the Spey snakes its way north past forests and fertile fields through the distillery town of Rothes to the coastal plain of Moray and its final destination – Spey Bay. Due north, Elgin, Lossiemouth and the Covesea lighthouse, a prominent white pillar that guides ships safely up and down the Moray Firth. Across the open sea, 100km (62 miles) due north from where we are standing, the distinctive snow-capped peak of Morven can be seen rising above the Flow Country peatlands, the highest point in Caithness. Spinning around 180 degrees and now looking south-east, we make out the prominent white shape of the Buck, the highest of the Cabrach Hills.

At 460m (1,509ft), we reach the destination of the peat carts, a large area of cut-away bog with patchy snow that clearly marks out the old cut marks and peat banks, the source of the distillery's peat before the floor maltings closed in 1964. This area is known as Baby's Hill, the final resting place of a young woman who took her own life in the late eighteenth century. The body of Barbara McIntosh was hidden away in a peat bog high on the slopes of Ben Rinnes because the burial of suicide victims on consecrated ground was not allowed by the church at the time.[18]

We are now walking up into a huge area of mountain peats that blanket large parts of the upper slopes of Ben Rinnes and cover all but the steepest gradients or the highest summits. The snowline of last night's fresh snow arrives at 550m (1,804ft) and we walk on across uneven peat moss towards the summit ridge, where snow is being whipped around in the strong wind. Alternating between hard ice, deep snow and crusted snow that breaks to knee height when you put your full weight on it, we walk higher and higher across the windy plateau above the source springs of the Scurran and Rowantree Burns. The winter snow has collected amongst the eroded peat hags and is being actively reshaped into deep dunes and linear ripples by the strong winds whipping across the summit ridge. In the winter light, large sandblasted granite tors, sugar-coated with ice, stand out like weird stratified snow sculptures.

We see tracks in the soft snow and occasionally get a glimpse of a white mountain hare darting across the slopes before stopping abruptly for a few seconds and then moving on. As we get higher and closer, the summit appears and disappears in and out of cloud and spindrift. Finally, we walk up a gentle slope of gravelly granite and at 840m (2,755ft) reach the tor that marks the summit of Ben Rinnes. After scratching away at a white crusting of ice, we find the toposcope that maps out the distilleries that encircle this giant lump of granite.

It's too cold and windy to linger long, but from time to time we get tantalising glimpses of the Cairngorm mountains to the south-west and the Spey valley to the north, before the cloud snuffs out the view and blots out the sun. On a clear day, Glenlivet Distillery would be visible, but not today. The winds that sweep across the summit of Ben Rinnes and down its slopes are legendary and with today's windchill lowering the temperature to around -15°C, our fingers quickly begin to freeze and the thought of a celebratory dram evaporates. All we want to do is go lower, find some shelter, defrost and drink hot soup. Late in the afternoon and partially defrosted, we walk back past the distillery of which Alfred Barnard wrote in late 1880s 'no more weird and desolate place could be chosen', followed by 'it was not difficult to imagine the loneliness of winter in these latitudes'.[19] He also met the brewer on his tour of the distillery, Mr A. Mackintosh, who lived in the house where we slept.

The name Ben Rinnes is derived from the Gaelic *Beinn roinn*, meaning 'Promontory Hill'. It is the most distinctive summit in the lower Spey Valley and in my wanderings I have observed it from the hills above Brora, distillery car parks and the bogs of the Faemussach and the Mannoch Hills. Around 400 million years ago, this plutonic granite sentinel punched its way through the old metamorphic schists of the Scottish Highlands and after rising up and being re-sculpted during successive Ice Ages, now forms this hard, upstanding crustal intrusion. People have lost their lives on Promontory Hill, in battle, in World War II bombers, in bad weather. More happily, it has hosted weddings, parties, hill races up and down to its rocky summit, and the surrounding glens were once the home of illicit stills fed by the burns and springs that issue forth from the mountain's flanks.

The last time I climbed Ben Rinnes was twenty years ago. We called this the Sunday morning 'hangover walk' and it was worth getting up for. After a good Saturday night at Jimmy's at Carron Lodge on the banks of the Spey just yards from the silent Imperial Distillery, we would all set off from the car park of Benrinnes Distillery. It was a way of clearing the head before a big meaty roast of a Sunday lunch and a slow drive back to Aberdeen. Age obviously doesn't dim my desire to walk upwards, but each year the words of William Young Sellar, Edinburgh University Professor of Humanity (1824–90), ring louder in my head:

> There is a geological problem that puzzles me a good deal; perhaps you can throw some light on it? How does it come about that the Scottish hills with which I am acquainted, are so much higher and steeper than they were thirty years ago?[20]

Covering an area of 50km^2, Ben Rinnes is quite simply the most important mountain watershed in Scotland for making whisky, and at its heart lies pure, soft spring water, peat water or moss water. Peat was once cut from the slopes of the mountain for the malt kilns at Benrinnes and Glenfarclas, its streams powered distillery waterwheels that drove mechanical devices like pumps and stirring rakes, and its granite was quarried to build castles and distilleries. Today, its importance is manifest in the names of the distilleries that directly tap into its ability to capture large amounts of moisture. Benrinnes, Glenallachie, Aberlour, Dailuaine, Allt A'Bhainne and Glenfarclas all directly depend on it for soft process and cooling water. Tributaries on its south-eastern flanks feed into the Dullan Water that joins the River Fiddich at Dufftown before flowing north and downhill towards the Spey past Craigellachie. The six working Dufftown distilleries all use the River Fiddich for cooling water, as did the closed distilleries of Parkmore, Convalmore and the demolished Pittyvaich.

The peat mosses that have formed on Ben Rinnes are a fine example of the mountain peats that blanket large parts of the Scottish Highlands. They form at these altitudes, and sometimes on surprisingly steep slopes, due to a combination of persistently high rainfall, low temperature and acidic bedrock. Although many have escaped the peat cutter, these are fragile systems, gashed and gullied and easily eroded by humans, sheep, deer, wind or rain. On average, it rains or snows on the summit of Ben Rinnes for 250 days a year and, like today, its summit is frequently covered in cloud. Its peat bogs act like a great absorbent sponge, soaking up the rain and melting snow that with time feed springs, replenish wells, trickle into small burns that become larger burns and eventually spill out into the wide Speyside glens below, where rivers like the Lour, Fiddich and Livet flow. At times of heavy rain when the peaty sponge reaches capacity, water bursts off the mountain in torrents targeting the land below with peaty brown floodwater and debris. It is a place of granite and peat, and quite possibly the place someone had in mind when they came up with the words 'the best whisky was made from water, which comes off granite through peat'. While others claim better whisky is made from water running 'off peat and over granite', at Ben Rinnes both occur, and it matters not.[21]

In 2020, Scotland's list of operational malt whisky distilleries reached 122, its highest number for seventy-five years, and the number continues to grow. Many of the new distilleries have sprung up in cities or in the rolling agricultural lands of Fife, but for an industry rooted in its Highland and Island past, how many still rely, or partially rely, on water derived from catchments containing important peat mosses? The answer is almost seventy – in the past, the number would have been much higher. For a country with 20 per cent of its land surface covered by peat, the fact that around 60 per cent of Scotland's distilleries have sprung up in locations close to areas of peatland is testament to the importance of this water-saturated, organic soil. It is a clear and transparent link to the past. Whisky smugglers chose places to make their whisky with 'the purest mountain streams, flowing over moss and peat'. They were pragmatic people, understood the importance of natural resources and recognised the role that peat played in providing a secure and plentiful supply of high-quality water, as well as fuel. Other ingredients – grain, yeast, coal – could all be brought in, but water could not. It had to be local, abundant, clean and available throughout the year. The old bothy distillers used water to steep their barley, mash the malt, cool the distillate in worms or rudimentary condensers and reduce spirit strength prior to casking. They also used it to dilute and disperse their effluents and wastes – sometimes leading to detection, fines and the destruction of their valuable equipment. The silent power of water could also be harnessed to transport their casks.

The whisky industry is now one of the largest users of water in Scotland and in 2014 used 49 million m³ of water.[22] While the figure appears staggering, the largest use of water (73 per cent) is for condensers and less commonly wormtubs. After it has done its job, it is returned to the watercourse – slightly warmer, but otherwise untainted. The remaining 27 per cent is retained during the whisky making process – a form of net water extraction, or loss, from the aquatic environment. Forty-nine million m³ is a large number, but to put it into some kind of context, it is equivalent to the average amount of water that flows out to sea from the mouth of the River Spey every eight and a half days. Loch Ness, Scotland's largest freshwater loch, contains 7,500 million m³ of water, enough to supply the Scotch whisky industry for 150 years.

An Experiment with Peaty Water - the Story of Craigduff and Glenisla

While most people agree that soft water, maybe with a peaty brown tinge, is ideal for making whisky, there is still some debate about whether it plays a role in enhancing its phenolic flavour profile. Denis Nichol, Laphroaig distillery manager from 1974–80, armed with a gas chromatograph and backing up his senses with science, believed the aromas of bog myrtle (peppermint) and yellow flag root (phenolics) that grew naturally in Laphroaig's peaty acidic water supply contributed to the taste of his whisky.[23] This of course matters most in places like Islay and the Scottish Highlands where peat is more abundant and played an important part in the history and DNA of the region's whisky. Maybe the best place to form an opinion is in a peat bog. Taste some clean, natural peat-strength water and compare it with the tap water you remembered to bring with you from home, or the expensive bottle of mineralised *aqua* you just bought from a local shop. Bog water tastes different to me, not peaty, but smooth, slightly bitter with a dry finish.

The peaty water debate has been simmering gently for more than 100 years. In 1913, J.A. Nettleton was already doubtful about the contribution made by peaty water to the taste of whisky distilled in the Highlands and Islands, writing, 'Mysterious influences are ascribed to the use of moss-water.'[24] In the 1920s and without the help of modern analytical techniques, the chemist Stuart Henderson Hastie went further: 'there is a deep-rooted belief that variations in [whisky] character are due to the water', but there is 'little real evidence to support this belief'.[25] The great whisky scribe Michael Jackson held a different view. Writing about Speyside water, he said, 'the clean, mountain water picks up enough of the region's peat to import the smokiness that is their particular characteristic'.[26] I came across this piece of smoky sales-speak written in 2019 by a marketing team of an unnamed distillery: 'the 12-year-old is unpeated and matured in ex-bourbon casks, although the natural peatiness of the water on the island of Mull does tend to add a hint of smokiness to its finish'. While science has debunked some of the myths around water, the debate lingers on. In my many days working and journeying across the peatlands of the northern hemisphere with occasional stops to rehydrate, I have never detected a trace of smoky or peaty aromas and flavours in bog water. Peat needs to be burnt to release its smoky, peaty aromas. Without doubt, there are strong sensory signals coming from the peat moss, but if you detect peatiness in the water, it is probably just another case of sensory persuasion, olfactory hallucination or even auto-suggestion.

Such is the importance of peaty flavours in single malt and blended Scotch whisky that in the 1970s, experiments were taking place in distilleries and laboratories to find new ways of enhancing the phenolic content of the spirit. Almost all involved attempts to optimise kilning with peat, but one experiment stands out from the rest. It focused on the role of peaty water and links the peat-covered island of Lewis in the Western Isles with an important whisky-making town in north-east Scotland. Peat dominates the flatlands of this Gaelic-speaking island and lies at the heart of much of its culture, history and folklore. Apart from the islanders' use of peat as a fuel, exploitation of Lewis's natural peatlands to generate employment and income for the local community has been attempted on several occasions, some with more success than others.

PEAT AND WHISKY

Back on our travels we disembarked from the ferry on a Friday night, driving through the dark, rain-washed streets of Stornoway before heading out across the moors to spend a week amongst the peat bogs on the northern end of the island. This is the most active peat-cutting area in the Western Isles and when we woke the next morning, the locals were still bringing in the peat and building their well-crafted peat stacks, many constructed with the herring bone pattern of Harris tweed. Each dark October night, we lit a glowing peat fire; in the morning, all that remained was a milky white, almost weightless ash, with the residual smell of smoked haddock. Sunshine and showers were the order of the day and rainbows followed us wherever we went. At night, the lighthouses of mainland Scotland flashed at us across the Minch and when the rain clouds parted, the dark night sky was dominated by the sweeping arch of the Milky Way, and on one night, a tantalising glimpse of the Aurora Borealis.

Called 'unusual' and a 'curiosity' by whisky pundits and writers, the origins of Craigduff and Glenisla single malt whisky are to me one of the most interesting and unique tales in the story of peat and whisky. It is a story that starts in a loch in the Western Isles, continues across the stormy waters of the Minch, round the northern tip of Scotland, south to a fishing village on the Moray coast, and ends with a road trip to the whisky-making town of Keith, a journey of more than 650km (200 miles). Tasting notes of the pair often refer to them being 'unlike ordinary peated whisky', 'pedagogical, totally unlike anything I've had, and quite fascinating', 'a bit artificial'.[27] Many of the most experienced noses and palates have even failed to detect peat at all. They are both enigmatic, mysterious and rare whiskies, firmly entrenched in whisky folklore. There has been much discussion focused on whether the whisky was made in Strathisla or Glen Keith Distillery. This is academic since both distilleries share the same owner and lie on opposite banks of the River Isla, where new-make spirit is piped from the stills of Strathisla to the cask-filling station at its sister distillery. Founded in 1957, Glen Keith Distillery came to be used as an experimental plant, firstly for Seagrams and then by Chivas Bros. The late Iain Henderson, who in 1989 became one of Laphroaig's most influential distillery managers, was in charge of the experiments that aimed to create an Islay-style whisky on the mainland for the Chivas family of blends. Not only was peated malt used, but peat-enriched water was specially brought to the distillery to enhance the phenolic content of the process water. These experiments were unique because the smoky peatiness of the spirit came not just from the kilning of malted barley but also from the manufacture of an unusual process water produced in a smoke plant on the shores of a loch.

This morning, I'm travelling south along the Arnish Moor and soon reach the shores of Loch Breugach, in Scots Gaelic literally 'the loch of the lie or falsehood'. Why it should have this name probably relates to the fact that we have just driven past one of the island's most notorious crime scenes.[28] In May 1964, John MacLeod and his brother Donald from Crossbost were cutting peats when two feet down their *tairsgeir* cracked into the skull of a human being. While the rest of the body was poorly preserved with the 'consistency of rubbery seaweed', the cadaver's clothes, hair and writing quills were in good order. The local constabulary was called out and after examination, the body was released and deemed to be of 'historical interest' only. Forensic science then took over and identified the body as one of a poor but scholarly young man who had been murdered in the early 1700s by a blow to the back of his skull. If that wasn't bad enough, he had his shoes nicked. The clothes of Arnish Moor Man may now be just an item in the collection of the National Museum of Scotland, but the story doesn't end there. Local tradition recounts a tale of two young men who went to Arnish Moor to collect grouse eggs. They fell out, one was assaulted with a blow to the head from a stone, killed and buried. The assailant quickly left Lewis and escaped for a life at sea, before making an error of judgement by returning to Stornoway where he was recognised, confessed to his crime, convicted and hanged as a murderer on Gallows Hill.

I pull off the road and park at the entrance to what looks like the footprint of an old industrial site. The locals know this place as the 'old Chivas Regal site' and I have come to meet historian Ken Galloway and George MacDonald, once a factor for the Stornoway Trust, which was gifted a large part of the island in 1923 by the then owner Lord Leverhulme. Ken, agile and wiry, sets off across the bog looking for the loch inlet pipe. George – moustached, well-dressed, with a big smile and an uncanny resemblance to Whyte & Mackay's retired master blender Richard Paterson – is more circumspect, absorbing his surroundings and rekindling past memories. In the tradition of the island's storytellers, George told me his tale of the place:

> I started with the Trust in 1974 or 1975, and to the best of my knowledge, the arrangement with Chivas Regal was already in place before I started. This site was used for several items of plant designed to burn peat and pass it through water piped from Loch Breugach, which was then stored on site and used to fill drums that were taken to Keith on the mainland, where the peaty water was processed for their whisky. There was a little concrete building on site that housed the operator, one room and a toilet. The man who worked there was called by everyone Iain Uig; his right name was John Macdonald, from Geshader in Uig. Iain stayed out there for long stints on his own and passed the time by taking long walks into town and through the castle grounds. After the plant ceased

to be used, Iain Uig was given a job with the Stornoway Trust and he joined the squad that worked from the sawmill in the castle grounds until he retired. He was a large, strong man who never touched a drop and slept in the sawmill on one of the benches. I think the boys in the grounds helped Iain cut the peats for burning at the plant and I remember seeing the peaty water, which understandably was very dark and oily and had a distinctively peat tar smell that was not as pleasant as peat smoke.

I had hoped to visit Iain Uig's place of work before now, but the global Covid pandemic and travel ban delayed my journey time and time again. I then received word from Lewis that his concrete home and place of work had been bulldozed. George later sent me a photograph taken by his late father in the early 1970s of the site in operation – a functional white building with windows, a door and an overhead power supply surrounded by fifteen to twenty green metal 45-gallon steel drums. Set apart from the building are four sets of large concrete tanks and smaller header tanks that stored the pre-process loch water. It's clear from the photograph that this was not some cobbled together experiment but a well-organised plant built to last and produce significant quantities of super-peated water. I asked George why Chivas chose to put the smoke plant here? 'Simple,' he replied, 'there was mains electricity, a water source, good peat, a main road and the port of Stornoway just a few miles away – logistically, it made complete sense.' I filled a bottle with Loch Breugach water. It was very soft, pleasant to taste with a light peaty colouration – ideal for a late evening dram.

As the three of us walked around the site, George pointed out the foundations of the water storage tanks, the remains of the power cables and the loading bay where the drums of super-peated water were rolled on to the backs of waiting lorries before being taken down to the harbour. Amongst the bulldozed remains of Iain's concrete home, I found old wiring, rusting metal remains of the plant, including pipes, valves and sidearms, as well as small wooden cask staves and a large piece of porcelain toilet. Outside the fenced compound, Ken took us on a short hike across the peat moss as we followed the line of the water inlet pipe down to the loch. On its shores, the buried pipe emerges from a peat bank into an old wooden pumping shed and then through a series of connectors into the loch, before disappearing from view into deep water. All around the plant are old peat banks, one of which is still in use – the same peat banks where Iain and his helpers had cut the peats for the smoke plant.

After I left the Isle of Lewis, I received an email from Sandy Matheson, former provost of Stornoway, who had heard about our expedition to the old Chivas Regal site. There was a new twist to the tale:

I was Chairman of the Trust at the time and in response to a request by Stewart McBain, the Operations Director of Chivas, we took one Rocky Munro, their Chief Engineer, over to Stornoway to discuss things. In due course, Chivas tried an experimental run on James Street and Keith Street corner, but on one occasion, a Saturday afternoon, the smoker had a blow back and consequently Rodishan Macleod, who had his home and office next door, decided that the whole thing was dangerous, so we and Rocky organised the smoke plant out on the site at Loch Breugach. We then received an invitation to stay for a few days at the distillery in Keith and were lodged in the house occupied by the Chivas directors on their frequent visits from Seagram, the USA parent company. We had a very detailed tour and inspection of the distillery, bonded warehouses and other facilities.

This was not the first time that the rich peat resources of the Isle of Lewis were used by the whisky industry. When the engineer Joseph Mitchell visited the island in 1838, Stornoway had its own large and short-lived distillery powered by local peat.[29] It was owned by Mackenzie of Seaforth, but when the teetotaller, prohibitionist and clearance landlord Sir James Matheson, 'The Opium King', acquired the island in 1844, he demolished the distillery on the Shoe Burn and in its place built Stornoway Castle and its various outbuildings. As early as the 1830s, there was a steady trade in cut distillery peat between the ports of Stornoway and Leith, and Isle of Lewis peat was being shipped down the west coast of Scotland to the maltings at Littlemill Distillery on the shores of the River Clyde.[2]

I couldn't leave the Isle of Lewis without exploring what remained of the island's greatest project to turn peat into profit. Just outside Stornoway a track up onto the moor leads to a large granite headstone that reads, 'This stone marks the site of The Lewis Chemical Works, set up by Sir James Matheson to extract paraffin from peat, 1857–1874'. The tale of the works, chronicled in Ali Whiteford's fine book *An Enormous Reckless Blunder*, is a story of a pioneering desire to generate employment and wealth in the Victorian industrial age on a remote Scottish island.[30] Matheson started burning and distilling peat to make refined chemical products such as paraffin oil, lubricants, wax, tar and sheep dip. Large quantities of peat were cut and transported along two tramways and a canal with lock gates to the purpose-built chemical works. Initially, all went well, but things soon started to go wrong. The works had problems of its own making – mass poisonings wiped out the local fish population and the people of Stornoway were frequently enveloped in a cloud of peat reek. There was an explosion and fire, and a combination of bad management and fraud ultimately led to financial ruin. The work's problems were compounded by a rapidly changing energy market – paraffin oil was now being produced on an industrial scale in Scotland's Central Belt, and when oil was struck below ground in Pennsylvania, the writing was on the wall for the Lewis Chemical Plant. The works are long gone, but artefacts from the plant have been repurposed as fence posts around the island and make appearances as unusual ornaments in the gardens of the residents of Stornoway. Locally made bright-red bricks and rusting rails are still common finds, and out on the moor to the north of the Creed River, the tracks of the tramway and the remains of the old peat canal are still visible. So are the scars of hundreds upon hundreds of parallel peat banks that fed the works for its short and tumultuous life. It was a huge, failed operation.

While peat from the Isle of Lewis has historical links with the whisky industry, it doesn't explain why Chivas specifically chose to return to Lewis in the 1970s when they could have built a smoke plant closer to home amongst the peatlands of Moray or Aberdeenshire. Maybe the answer lies in the quality of the peats on the island. They are unusually pure, comprising rough grasses, bog mosses and shrubs. A lack of tree stumps poking out of peat banks can be explained by the fact its bogs have been treeless for the last 4,500 years. The Chivas operation was built on one of the largest expanses of blanket bog in Europe, and if the production process had been embraced by Chivas, the outlook for the peat moors of Lewis would have been bleak. Craigduff or Glenisla whisky was never officially released by Chivas as a single malt, the intention being to use the super-peated whisky for blending purposes. Both malts made a belated and brief appearance in a 1990s expression called the Chivas Century of Malts, a showpiece blended malt containing whisky from 100 different distilleries. Fortunately, both Craigduff and Glenisla were preserved as single malts by the independent bottler Signatory, rare and fascinating examples of distillates produced in a completely unique way. Andrew Symington, the owner of Signatory, continues the story:

Lightly peated barley from Glen Keith maltings was used in conjunction with controlled amounts of concentrated peated water being added to each wash charge. Peated water was brought in 45-gallon drums from Stornoway, on fishing boats into the port of Buckie. The peated water was run through the small still at Glen Keith, which was coupled to an angled condenser and water driven off to concentrate the peatiness in the remaining water. It is understood that ten gallons of the concentrated peated water was added to each wash charge.[31]

The first experiment at Glen Keith ended on 4th April 1973 and Signatory have bottled six Craigduff single cask whiskies from this spirit run, ranging in age from thirty-two to forty-five years. More than four years later on 7th July 1977 (07/07/1977) another experiment was successfully completed and the results bottled as Glenisla. To date, Signatory have bottled it nine times as a single malt; coincidentally the 34-year-old was bottled on 11th November 2011 (11/11/2011). I have been lucky enough to sample these whiskies. These are my tasting notes and they hint at a further twist in the tale:

SAMPLE ONE:
Craigduff 1973/2005, 32 years old, sherry butt, 53.9%, single cask #2514.
Colour – golden.
Nose – lovely, sweet first-aid plasters, spicy and peppery, distinctly smoky.
Taste – malty and fruity, but slightly dry and bitter.
Finish – lingering and very pleasant. This could easily be an Islay-style peated, sherried whisky. A success.

SAMPLE TWO:

Glenisla 1977/2006, 28 years old, hogshead, 48.8%, single cask #19601.

Colour – light and winey.

Nose – unpleasant, pungent and aromatic, a collision of volatiles, glue, stale leather-bound books, old wood, ink, weird varnish, a closed-up decaying office.

Taste – apple juice and pears, sweet and sour, stagnant water, maybe briny, flat, uniform and thin, not unlike a grain whisky, no smoke.

Finish – it just ends, instantly. A bizarre whisky with a nose like no other. Unless you are into weird whisky, this was less than successful.

The fact that the results of these experiments can be tasted after such a lengthy maturation period is remarkable and a tribute to Signatory, but the two single malts tasted transparently different to me. On my return from the Western Isles, I contacted Alan Winchester, then master distiller at the Glenlivet Distillery, to ask if he knew more. These are his words:

Glenisla was made at Glen Keith from the peaty water produced in Lewis, Craigduff was made from peated malt which was purchased from a commercial maltster who received it from Glen Garioch Distillery...When I went to Glen Keith in the mid-1980s, we still had a stock of the peated water that was made by lighting a peat fire and pulling the smoke through a water tank to catch the phenolic essence. It had a strong phenolic smell and if you got any on your hands, it remained with you after extensive washing. The last drums were dispatched to New Zealand Wilson's Distillery following its acquisition by the Seagram Company Ltd in 1981.

The two experiments that resulted in Craigduff and Glenisla were conducted to compare single malt whisky produced in a traditional way using peated malt, with a new method by infusing water in the still with an essence of peat. I could sense science at work here. In the 1970s, Chivas Bros were aiming to create their own peated whisky for their blends, and driving this was operations director Stewart McBain who was well-known in promoting a more scientific approach to whisky making at Chivas.[32] The production of the super-peated water was done in incremental stages, starting with peaty loch water from Lewis, infusing it with smoke produced by burning local peat, and then concentrating the phenolics still further in the stills at Keith. The old Chivas Regal plant produced the most expensive process water ever used in the history of whisky making, and it was in response to the increasing threat of drought on Islay. And remarkably, the final drums of Iain Uig's peat-infused Loch Breugach water travelled across the world to make whisky in New Zealand. And there the tale runs dry, for now.

'Drought turns the whisky stills dry.'

Just below a story about Richard Nixon's new running mate in the autumn US elections ran the *Daily Mirror* headline 'Drought – Whisky and Water'.[33] The big news was that the rains had finally arrived on Islay and the drought that had closed Port Ellen Distillery and threatened several others in 'Little, Lonely Islay', was officially over. According to the paper's reporter, 'the morale of the islanders, and all whisky drinkers, rose towards the heavy, sodden clouds'.

Forty years later, another piece, this time in the *London Observer*, 'Drought turns the whisky stills dry.' Half the ten distilleries on the Isle of Islay have been hit. A layer of peat at the highest point of the island has become dehydrated, causing several burns to stop flowing. This in turn means that the lochs the distilleries draw from are not topped up and are quickly emptied.' An unusually dry and warm spring in the Inner Hebrides in 2008 was the cause of the water shortage. History was repeating itself once again.[34]

Distilling in the Highlands and Islands of Scotland was traditionally carried out between the months of October and May. The summer was the season when peat was harvested from the mosses and the barley crop brought in from the fields. The men were away from the distilleries at a time that often coincided with water scarcity in the warm, dry months. Cold water, not warm water, was needed for the condensers – surface water could also become stale and tainted. As distilleries grew in size and the supply of raw materials became less seasonal, the quiet season shortened from sixteen to twenty weeks to just two to three, and the demand for water increased. Even though the whisky industry is well-known for its water use efficiency and has been particularly innovative, expansion and the building of new distilleries has increased pressure on local water supplies. Distilleries like Talisker, Glen Garioch and Clynelish have all taken the step of bringing on stream completely new water sources to improve supply security and future-proof their production. Caol Ila ingeniously uses cold seawater from the Sound of Islay to cool its process water, neatly sidestepping any seasonal issues of water temperature or availability.

Despite Scotland being, in the minds of many, synonymous with the three words cold, wind and rain, drought is not uncommon and its frequency may be increasing. In modern times, the years 2018, 2008, 1980 and 1976 are all remembered in the Highlands and Islands for their water shortages and temporary closure of distilleries. 1968 was the year when the drought in Islay and Skye made several companies, including Chivas Bros, rethink their dependence on the two islands for heavily peated malt for their blends. While Chivas was experimenting with super-peated water from Lewis, the Brora Distillery was brought back to life, and from the early 1970s started to produce a heavily peated malt to meet the demands of the blenders. Its maltings were already closed and the peat shed redundant, but Glen Ord Maltings stepped in to become the new supplier of peated malt. Other companies were doing the same thing for the same reason – in Campbeltown, Springbank Distillery stopped buying peated Islay whisky for its blends and started making a super-peated single malt called Longrow. On the Isle of Mull, Tobermory Distillery created Ledaig.

The countrywide drought of 1976 is better remembered on the UK mainland for its burning forests, fire-scorched heathlands and hosepipe bans, but distilleries were suffering as well. The Glen Grant Burn, supplier of water to Glen Grant and Caperdonich distilleries, ran dry for the first time in 136 years.[35] Nearby Dallas Dhu also lost its water supply in the drought of the same year.[36] Seasonal water scarcity has continued into the twenty-first century. Glenfarclas, situated in the shadow of Ben Rinnes, has to periodically reduce production to cope with summer water shortages from its mountain springs and peat-fed streams. Balmenach Distillery in Speyside also has to manage periods of low water supply. Warmer winters resulting in less winter snow cover may be to blame. In dry years, as Pulteney Distillery found in 2021, you simply make less whisky.[37]

Inevitably, famine turns to plenty and Scotland's weather has a habit of dramatically reasserting itself. Peat is notoriously difficult to wet up after it has gone through a lengthy period of drying – it becomes hydrophobic, literally 'water hating'. When rain finally arrives, it may simply run off the bog surface, and drought quickly turns to flood. North-east Scotland, in particular the lands of the Spey, Moray and Aberdeenshire, is well known for its important peat-rich watersheds, and almost 200 years ago one flood stands out as being the most revered and talked about in Scottish history. It took place in a part of the world where peatland, water and whisky all feature in this remarkable story of an extreme hydrological moment.

Biblical Storms and Modern-Day Floods

On the third day of August 1829, 'outbursts of subterranean water' issued forth from the mountains, accompanied by earthquake shocks, a hurricane, thunder and lightning. After an unusually long and severe drought, 100mm or more of rain fell from the skies on the lands of north-east Scotland in just twenty-four hours. The result of this was a catastrophic summer flood of almost biblical proportions that in time became known locally as the 'Muckle Spate'. It helped inspire one of Sir Edwin Landseer's most famous paintings, *Flood in the Highlands* (1860), in which a terrified family with all their possessions are pictured huddled together on high ground with their dogs and sheep as the tempest rages around them.

Stories of the impact of this great flood were famously recorded by a local landowner, Sir Thomas Dick Lauder, who set out from his ruined estate to tour and record the devastation of the region.[38] Many of his stories are colourful – sheep 'were found alive on the tops of trees at the foot of the garden', and my favourite, 'the current…actually carried off the gardener on one of his melon frames, to take an aquatic excursion among his gooseberry bushes and cauliflowers'. There is much vivid detail in his observations and he describes the scenes of devastation he came across along the banks of the Spey, Nairn, Findhorn, Dee and Don. Water in the lochs and rivers reached levels never previously recorded – bridges collapsed and were washed away, roads damaged, crop and grazing land stripped bare of soil, properties inundated and livelihoods wrecked. He wrote how 'the mill and brewery at Keith were saved by the locals, but there was much destruction of the vats, casks and beer, and the whole winter stock of coals was carried away'. Scotland's new Highland infrastructure was severely damaged and the engineers who had built the roads and bridges had to return to rebuild and repair the havoc.

The timing of the flood was significant for some of the Highland's new distilleries. It occurred just six years after the Excise Act of 1823, a moment that had triggered a wave of construction resulting in a new high in the number of legal distilleries operating in Scotland. Several notable names were destroyed by the Muckle Spate, including Lochnagar Distillery. Originally built on the north banks of the River Dee in 1823, it was later rebuilt on higher ground to the south of the river in 1845 by John Begg.[39] Overnight rivers in Speyside rose from three to 17ft above their normal levels and flood water breached the walls and filled the vaults of Ballindalloch Castle. The Tamdhu Burn washed away cottages and a carding mill – the distillery would have gone as well, had it been built. The Spey itself washed away part of the glebe lands of the Macallan estate[35] and close by, water cascaded off the slopes of the Conval Hills and the peat-rich flanks of Ben Rinnes into the Lour Burn, over the spectacular Linn Falls, on past the Aberlour Distillery and down through the Spey valley. Nearby, the new Benrinnes Distillery was not so lucky, lasting just three years after being built close to Whitehouse Farm. The deluge swept down from Promontory Hill on past the doors of the distillery, taking the building and its equipment with it.[40] Looking at the Lowing Burn today, it is hard to believe that such a small stream could turn into such a destructive force, but close to the farm its incised channel forced water through a narrow gulley and created the power to destroy a distillery. A new site on higher ground was chosen one kilometre away, and although the distillery has been rebuilt and enlarged several times, the modern distillery still stands.

For all sorts of reasons, rivers were a magnet for the location of distilleries, and with it came the threat of flooding. One of the most catastrophic floods occurred in 1883 when the Kentucky River in the USA rose 50ft above its low-water mark and inundated three bourbon distilleries, including one owned by E.H. Taylor Jr.[41] For obvious reasons, flat pieces of land next to navigable rivers, canals or ports areas are attractive and convenient places to store whisky. In 1985, a bonded warehouse complex known locally as 'The Long John' in the East End of Glasgow, owned at the time by Allied Distillers, was inundated by flood water from a tributary of the River Clyde. Attempts to alleviate the flooding went badly wrong and empty whisky barrels were buoyantly carried away in the spate towards the River Clyde. Fortunately, the Clyde Port Authority was on hand to capture and retrieve the empties with the help of a boom.

Flooding has always been an occupational hazard for many distilleries and their warehouses, built out of necessity on low-lying land or on the banks of fast-flowing streams and rivers. 'Today's rain is tomorrow's whisky' is a well-used Scottish proverb, but when it rains intensively and incessantly, today's or last week's rain can wreak havoc on the defences of a distillery. In my experience, Scotland does not have a predictable wet season, or for that matter a predictable dry season, and flooding can occur at any time of year. Flood barriers and control measures are very much part of the operating procedures of modern distilleries, particularly if they are built on an active floodplain. Look no further than Glen

Moray on the River Lossie, and the Borders Distillery in Hawick, located just a few metres from the River Teviot, one of Scotland's most flood-prone rivers.

Even in the twenty-first century, and despite flood management plans and sophisticated defences, flooding continues to be a major issue for some of Scotland's distilleries. The Glen Moray Distillery flood of 2002 is depicted on a whisky bottle label, with a running warehouseman in pursuit of his barrels shouting, 'not again' as the River Lossie floats his casks away downstream. A chest-high plaque by a warehouse door marks the flood water level that day on 16th November 2002. It happened again in 2009 and it will not be the last time.

January 2016 is remembered in many parts of the UK and Ireland for the misery it brought as heavy rain fell on saturated ground that had previously suffered month after month of persistent rainfall. It was one of the wettest periods in living memory and led to flooding across the whole of the British Isles and Ireland. Midleton in County Cork and Bushmills on the Antrim coast, as well as Glencadam and Edradour, were affected. In October 2017, the River Isla in Keith again burst its banks and three employees, who became trapped in the Strathisla Distillery, had to be rescued by boat. A few years earlier, a flash flood on the same river triggered another aquatic rescue when water overran the Strathmill Distillery and the manager and two operators were rescued from the still house.[42]

Unless you are a hydrologist or, whisper it, an environmental protection officer, your focus when you visit a distillery will probably not be on the contours of the land, the proximity of the warehouses to watercourses, or the whereabouts of local peat bogs. If it's raining heavily, maybe it should be, since the peatland water reservoir has an important role to play in downstream flooding and flood mitigation. The way in which a peatland is managed and restored will influence its capacity to store water, although how this changes run-off intensity and flooding is still a surprisingly controversial subject. Some scientists question whether the sponge analogy for peatlands is really helpful at all, suggesting it oversimplifies and misrepresents their long-term capacity to store water and slowly release it. Like peats, sponges are highly porous and generally soak up water quickly, but unlike peats, they lose it just as quickly when squeezed. Peats tend to hold onto water longer and don't drain as fast as sponges. Some have much smaller pores, absorbing water and holding onto it even longer. Peatlands are not all the same and it would be a mistake to think that peat has an infinite capacity to soak up rain and remove the risk of downstream flooding.

The simple fact that peat consists of 90 per cent water means that it has an enormous capacity to both retain water and release it. Bogs also have a natural annual cycle of seasonal breathing, swelling during the wet winter months and shrinking during the dry summer.[43] This huge natural reservoir provides a means of not only soaking up precipitation but also slowing down runoff and reducing the severity of flooding – 'flattening the curve', a hydrological term, but very apposite at the time I write these words. A peat moss, degraded, thinned by erosion and losing its blanket of *Sphagnum*, provides little flood protection to communities and businesses downstream. A functioning, growing moss has the potential to provide a steady supply of water to a downstream distillery and ameliorate the risk of flooding. In the words of W.H. Pearsall, one of our most eminent twentieth-century ecologists, 'When all is said and done, peat has a great virtue not often realised: that of storing water in a catchment area and releasing it continuously long after rain and floods have ceased'.[44]

In the past, many of the UK's peatlands were drained systematically to lower the water table and 'improve' the land for use as grouse moors, grazing land or to plant trees. As is all too evident in places like the Flow Country, the end result was often a degraded, de-vegetated landscape, a drier and thinner peat in which water movement downslope was speeded up by ploughing deep, parallel furrows. The controversy about flooding centres around the deliberate blocking of these old, man-made drains, a popular restoration strategy that aims to recreate the pristine ecosystem of the past. Some argue that this creates a different and new type of flood risk since a thin and dry covering of peat with a lower water table is a more effective sponge, compared to a saturated bog full of water. While all peatlands are different, researchers and land practitioners are generally happy to conclude that restoration moderates flashy rainfall events by creating a natural complexity in the peatland landscape of pools, diverse vegetation communities and micro-topography that slows down water movement. Damaged, dry peatland does not hold water well, particularly if it becomes hydrophobic, and has little capacity to dampen a major storm surge. In contrast, a well-managed, restored, thick peat will hold onto water and buffer the hydrological system, lessening the risk of flooding and the danger of drought. In many ways, the impact of the degradation and thinning of this veneer of peat is similar to the effects of deforestation. Tree removal often leads to faster run-off, soil erosion and potentially catastrophic flooding. Examples from across the world, both past and present, demonstrate this clearly – the Rhine Valley, the Spey, the Amazon rainforest. Like peat, forests act like a natural sponge, holding soil, water and snow in place, soaking up rainfall and releasing it slowly over time.

Flooding of course occurs in catchments both with or without peat, but there is no question about the importance of peat as a natural water reservoir. It has been estimated that 10 per cent of the world's freshwater is stored in its peatlands and in the UK, 70 per cent of our drinking water comes from upland catchments, many of which contain important peat mosses. To put it simply, degraded and damaged peatland will store and supply less water, but unlike the outflow of a reservoir, they cannot be regulated or quickly turned on or off at times of heavy rain or water scarcity. Every peat bog has its limits and they need to be managed well over the long term to maximise their capacity to store water. But like drought, there is a growing awareness that the frequency of flooding in countries like Scotland is increasing, and the finger is being pointed firmly in the direction of global environmental change.

So apart from scale, very little has changed in modern times, and a large proportion of Scotland's current distilleries are as dependent on peat mosses for their water as were the distillers of the past. Out of necessity, our whisky-making ancestors used water frugally and as distilleries grew in size and number, security of supply and purity have become more and more important. While peat bogs may be hidden from view, lurking in the clouds and mist of the surrounding hills, they are central to the operation of many of Scotland's and Ireland's most famous distilleries. In the words of Jim Murray, 'without water there would be no whisky'.[45] The whisky industry needs to take care of its peat bogs, in the same way it cares for its water supply. Without peat bogs, many distilleries would never have existed.

Eight

Peat and Whisky – A New Age

'It's an interesting thought that the heather growing purple
out on the moors today will be giving its flavour to
Johnnie Walker 3,000 years from now.
And that men will still be cutting the peat.'[1]

Irish - a Whiskey Reborn

The story of peat and whisky could have been about Ireland.
Accompanied by my brother, the paramedic, we have arrived at
a time when there is much talk of a new and deadly virus in the
air. Leaving Dublin and driving west, we sense a growing unease,
a strange tension in this normally relaxed, confident country. We
enter the Midlands of Ireland, a flatland of bogs, rivers, canals, small
farms and towns, churches and ancient monasteries, and make our
first stop on the Old Dublin Road in Monasterevan – busy, dusty
and tired looking on a Saturday afternoon. Tall brick chimneys,
broken roofs, street-side warehouse doors and windows all painted
company maroon red are all that remain of Cassidy's Distillery,
closed in 1921 after 137 years of distilling. At one time, it used
peat to directly heat its stills and dry its grain. In the 1800s, it even
became a source of currency as Mr Cassidy minted 'turf tokens' to
be used in exchange for carts of peat brought in by local turf cutters,
who were suspicious of coins and preferred a token to exchange for
goods.[2] Monasterevan was one of twelve distilleries that once operated in the peat-rich Midlands; now there are just two.

While the Scotch whisky industry has crested successive economic waves and wallowed in periodic depression, the story of
Irish whiskey is quite different. It once dominated the world before

a catastrophic, almost terminal collapse brought it to its knees. It is a story that has been well told by others, but what fascinated me was why such a famous whiskey-producing country, containing 1.2 million hectares of bog, only surpassed in Europe by Finland in terms of bogginess, is associated with a soft, sweet, mellow style of whiskey without a hint of peat? Scotland is a country blessed with huge fossil fuel resources; Ireland is not. So, what led Ireland to import coal for its distilleries when a much cheaper, local fuel was on its doorstep? Why did Irish whiskey take a different path away from the peat smoke aromas that became the very essence of Scotch whisky?

At one time, Irish whiskey was probably as peated as Scotch. Peter Mulryan, author of *The Whiskeys of Ireland* and a founder of Blackwater Distillery, told me, 'The *poitín* makers used the bog for many things; soaking the grain in bog water, drying the malt and fuelling the small stills.' In eighteenth-century Ireland, there were essentially two types of whiskey produced: legal, taxed parliament whiskey made in big cities like Dublin and Cork, and illegal *poitín*, literally 'little pot', made in poor, rural Ireland. A complex and misguided new tax in 1779 levied on the number and size of stills backfired badly for many legal distilleries and put them out of business, leaving the door wide open to fiery, peaty *poitín*. When in 1823 the government replaced the tax on stills with a smarter piece of legislation taxing the amount and strength of spirit produced, the industry was back in business. Illegal distilleries turned legit, began to grow in size, and the small stills of the *poitín* makers were marginalised.

I asked Peter why peat didn't continue to be a part of the story of Irish whiskey: 'In two words, industrialisation and scale. Expansion of the industry needed more energy to heat the larger stills and peat wouldn't do it, coal would.' Still sizes became bigger and distillers turned to the higher calorific value of coal. Distilleries like

Cassidy's at Monasterevan and Daly's at Tullamore grew larger and were next to navigable waterways that brought barges full of coal from the coast. In the nineteenth century, Irish pot stills were often two or three times the size of their Scottish counterparts, and the need for coal meant that new distilleries sprung up in coastal cities with easy access to imported coal. There was no more potent a symbol of urban industrialisation than the changing Dublin sky-line that became a forest of brick chimneys belching coal smoke from distillery fires. The tall chimneys were not built to improve air quality for the people of Dublin City or the distillery's workers but to increase the updraft from the coal furnaces. The result – more steam and more power.

But there is more to the story of why Irish whiskey took a turn away from peat. Carol Quinn, Head of Archives at Irish Distillers, is clear about how a geographical and social divide in Ireland became a whiskey divide:

> Scottish illicit whisky was peated and drunk by both the crofters and the rich lairds, it straddled the class divide. As Scotland's rural distilleries flourished and markets expanded, many retained their links to the peated style. In Ireland, peated *poitín* was largely consumed in remote rural areas – there were no middlemen to bridge the gap between the cities and rural Ireland. As a result, the rougher illicit Irish whiskey didn't find its way too easily into cities like Dublin and it never became a drink of the middle and upper classes.

When the coal-burning city distilleries expanded in size, the house style of Irish whiskey had already been set. By the end of the nineteenth century, the Thomas Street Distillery in Dublin had become the largest pot still distillery in the UK with an annual production of nine million litres of whiskey – huge even by modern-day standards. Apart from a period during World War II when coal became scarce, peat never again played a significant part in the manufacture of Irish whiskey.

Irish whiskey also had a tradition of drying malted barley in a closed kiln. It is not known when this started, but the net result was an untainted malt devoid of any peat smoke influence. When Barnard visited Ireland in the mid-1880s, closed kilning was already standard practice. He makes no mention of peat, writing in his own effusive style about 'extensive coal sheds'. Carol Quinn again: 'When coal started to be used for the drying of malt instead of peat, they closed the kilns because coal smoke is unpleasant', and rightly points out that peat could disguise the subtle, sweeter flavours of Irish whiskey. Traditional Irish pot-still whiskey uses 30–50 per cent unmalted barley in its mash bill, so any influence of peated malt would have always been naturally diluted in the new-make spirit.

After the Excise Act of 1823, legal distilleries went from strength to strength and by the end of the nineteenth century, Irish whiskey had become the largest player on the global stage. In the late 1800s, Dublin-made whiskey was traded across the world and had an unequalled reputation in terms of quality. It had many high-profile advocates, including the cricketeer W.G. Grace who appropriately took a whiskey highball, a double whiskey and soda, both during lunch and at close of play.[3] But already there were signs of a change in the wind and a challenge to the dominant position of Irish whiskey. An advert for 'Caledonian Peat Reek Whiskey' placed by a grocer and spirit dealer in Dublin[4] now seems strangely prescient and hints of the change to come: 'This very superior spirit stands unrivalled in this country', and 'has a richness of flavour peculiar to itself'. The big four Dublin distilleries at last began to take the threat seriously and in 1878 sponsored a book, *Truths about Whisky*. In it, they highlighted what they saw as the confusing and dubious practice of mixing Scotch with small amounts of single pot-still whiskey and marketing it as pure pot still. They had little time for grain whisky, dismissing it as 'Silent Spirit', and blended Scotch whisky was 'adulterated' by grain whisky, or 'Sham Whisky'.[5]

What then followed was a dramatic shift in the centre of gravity in the whisky world triggered by American prohibition that created an opportunity for Scotch whisky built on the back of blending. Carol Quinn again: 'The Irish whiskey establishment, particularly the Dublin whiskey makers, were committed to their product, single pot still, and wouldn't blend it with the product of the Coffey still. During American prohibition, they wouldn't break the rules and sell to middlemen', the bootleggers. Scottish distillers had no such quibbles, so Scotch whisky found its way to the USA and the Scotch style and brand expanded and put down roots. When prohibition ended, Scotch was ready to go, and the Irish were nowhere. It was a spectacular fall from grace, triggered by the unwillingness of the Dublin distilling hierarchy to embrace change.

Carol talked about other factors that began to contribute to the demise of Irish whiskey:

For an export-driven product, World War I was a disaster. The targeting of merchant shipping effectively stopped the export trade until the war's end. Equally disastrous was the Anglo-Irish trade war of the 1930s that saw hitherto lucrative markets like India, Australia, Canada – all associated with the British Empire – closed to Irish products; Ireland and its output were effectively frozen out of the British Empire. Irish whiskey sales collapsed, with the final blow for many distilleries being World War II. Irish whiskey was being hit by the perfect storm and if external factors weren't enough, the Pioneer Total Abstinence Movement founded in 1898 saw one million Irish men and women 'take the pledge' and embark on an alcohol-free life.

PEAT AND WHISKY

Scotch distillers owe a large debt to the actions of the British Prime Minister Winston Churchill during World War II. Keenly aware that Britain would need products for export at the war's end, he protected the valuable barley crop from being used for the war effort and negotiated lucrative trade deals with the US, which agreed to minimum importation levels. Ireland, a neutral country, was entitled to no such war reparations. Irish distilleries fell silent – there was no export market and although the whiskey was of high quality, it was too expensive for the Irish people. Whiskey in Ireland was an upper-class drink, consumed by the establishment and unaffordable for the masses; in Scotland, whisky was traditionally an everyman's drink.

For all sorts of reasons, boom turned to bust and by the 1960s, only three distilleries remained in operation from a high of thirty when the industry was at its peak. As we leave Monastervin behind, our journey to explore the connections between Irish whiskey and peat now takes us into the Bog of Allen, or the fragments of what now remains of the *Móin Alúine* wetland that once covered more than 1,000km² of the lowlands of Ireland and ranged across nine of its counties. It started forming 10,000 years ago, and since the 1600s was drained and cut for fuel by hand and later on harvested at an industrial scale by machine. We drive along straight bumpy bog roads, cresting an occasional rise that marks the crossing point of an old narrow-gauge peat railway. I was surprised by the number of people that live out here in apparently randomly scattered homes. We drove straight through a place marked on the map as Bogtown without seeing a town, dwelling or soul. We pass churches, church schools and convents dotted haphazardly around the boglands. The small towns were a hive of activity, just one week before St Patrick's Day. Big modern houses, with long wide drives, white walls, cut lawns and hedges spoke of wealth and status.

We have decided to travel westwards and follow the route of the Grand Canal to the Shannon River. The canal opened in 1804 after taking almost fifty years to complete, becoming Ireland's great artery connecting the Irish Sea in the east to the Atlantic Ocean in the west. In the canal's path was the Tullamore or Daly's Distillery, with its huge yard that could hold 5,000 tonnes of coal shipped up the Grand Canal from Dublin. The old distillery, which closed in 1954, had a giant waterwheel powered by the River Condagh that flowed through and under the building. The town of Tullamore was once the terminus for the canal before the traverse of the country was completed and the canal extended to Shannon harbour. This man-made waterway connected the Midlands of Ireland to Dublin and the coast, facilitating the two-way movement of goods and raw materials.

Modern-day Tullamore has seen the return of whiskey making to this part of the Midlands and in the context of the rebirth of Irish whiskey, it is a significant event. William Grant & Sons Ltd acquired the Tullamore Dew brand in 2010 and have now built a large-scale distillery on the outskirts of town. The new Tullamore Distillery was constructed in the middle of a large peat bog as there was nowhere else to go. Not an easy thing to do, but the builders put it on stilts and sunk it deep into the bog. Although the core range is unpeated, this well-established company has made a major statement of intent that shows confidence in the future of Irish whiskey.

Along a branch of the Grand Canal is the Kilbeggan Distillery, quite possibly the oldest distillery in the world. Built on the banks of the Brusna River in 1757, with its square red-brick chimney and the words Locke's Whiskey painted in large white letters on the side, it was the first distillery in Ireland to receive a licence. It still has its huge working waterwheel, testament to the importance of waterpower in the past. In the Victorian age, it added a steam engine and boilers that were fuelled with coal barged inland from Dublin. We visited on a peaceful Sunday morning, the site a mix of museum pieces and the modern reborn Kilbeggan Distillery. Locke's originally had three small squat pot stills fired by peat and then by coal, but it never produced a peated whiskey. When distilleries were closing all over the Midlands only one, Kilbeggan, kept its licence and reopened in 2007.

The following day in a sea of Irish rain, we left the peat bogs and headed south to Blackwater Distillery, entering a completely different landscape; a place of rocks, hills, gradients, woodlands and castles. I wanted to finally meet Peter Mulryan, who somebody in the industry had described to me as 'quite a guy'. He is – dynamic, anti-establishment, a man who wants to shake up the Irish whiskey category, a man in a hurry and someone fond of the F-word.

Blackwater Distillery is unusual. It has a tactile dog called Shadow, a sofa in the stillroom, a big 'we have nothing to hide' shop-front window, a violin resting on a tank of new-make spirit, two Italian-made copper stills and big sacks of assorted types of grain piled high in a corner. I wanted to talk about peat, but Peter wanted to talk about grain. 'In Irish whiskey, peat is used as one of a range of flavour options, but most of what makes Irish whiskey distinctly Irish is grain and the use of oats, rye, unmalted barley, corn.'

I asked him if peat was ever likely to become an important part of the future of Irish whiskey. 'No,' he said firmly, 'the future is grain, different mash bills with different combinations of rye, oats and wheat – that is the "old-fashioned way", but the new world for distilleries like ours.' John Wilcox, self-styled 'Distillasauras Rex', is the master distiller – an artist by training, he worked across the US in craft brewing and distilling before coming to Ireland in 2018 and joining Peter's team. His passion is playing around with grain, but recently he tells me he has run the distillery's first batch of peated malt that he wants to call Peat the Magic Dragon. John is a larger-than-life character who is excited about what is happening in the new melting pot of Irish craft distilling. Although some distilleries have a vision to make peated single malt again and turn the clock back to a time when grain was kilned with peat cut from the bogs of Ireland, grain is the name of the game at Blackwater.

As a parting shot, I asked Peter what was behind the rebirth of distilling in Ireland? 'Economics and craft have triggered the current boom in Irish whiskey. The Irish economy has come out of recession and taking its lead from the craft brewing industry, new distilleries are popping up everywhere. They are all doing their own thing, not really joined up. Some will survive and flourish, others will fail.'

The following day, we reached the coastal city of Cork and on a visit to Jameson's Midleton Distillery, I asked Carol Quinn the same question. She had a different answer: 'In two words, Pernod Ricard. The 1988 merger with Irish Distillers, makers of Jameson, provided a route to market for Irish whiskey and opened up new export markets throughout Europe and the world. The subsequent success of Jameson was the stimulus that led to the current rebirth of Irish whiskey.' There were no handshakes at Jameson's, just a warm welcome. The deadly new virus had reached the shores of Ireland and we had all been told to be careful.

Jameson's Distillery exudes confidence in the birth of a new era for Irish whiskey. It is not only one of the most modern distilleries in the world, with three 75,000-litre pot stills and three column stills, it has a 'Method and Madness' experimental micro-distillery with three steam-heated pot stills, small enough in the past to be directly fired by peat. But the one word that stuck in my mind after setting eyes on the old Midleton Distillery, now the Jameson Midleton Brand Home, was 'scale'. It possessed a colossal waterwheel to turn machinery, a six-floored malthouse, a huge hexagonal red-brick chimney and the largest pot still in the world – a gleaming golden pot of 144,000-litres capacity made in 1825 from riveted sheets of copper. It is not only an object of great beauty, but one that exudes great ambition and confidence in the future, a symbol of an industry at its prime. It was so large that it had to be assembled on site and then the walls of the stillhouse built around it. It was fuelled by the best coal in the world – anthracite from the South Wales coalfield that came in ships to the quayside at Midleton. Shovelled by hand at a rate of four tonnes a day into the furnace beneath this copper monster, it would have taken six times the amount of peat – twenty-four tonnes – to generate the same amount of energy.

Before we left Ireland, we drove north from Cork along the cliffs and round the bays of the wild, wet Atlantic coast of the counties of Kerry and Clare, and later that day dropped down into Galway. I was keen to taste peated Irish whiskey before we left and that evening dived into Garavan's, the best whiskey bar in town. There is not a lot of it around and lightly peated Bushmills is usually the first name that comes to mind. There are occasional glimpses of it in the past like Hewitt's Whiskey, a rare peated whiskey first launched by the Cork Distilleries Company in 1960 and discontinued in 2004. After warming up with a Green Spot, much to the barman's disgust I chose cask strength Connemara Turf Mor – about as un-Irish a whiskey as you can choose, but a good way to finish my peat and whiskey pilgrimage to Ireland. In an effort to keep alive a peated style of whisky that had all but disappeared, John Teeling in the 1990s created Connemara, a Scottish style of whisky. Double distilled using Scottish peated malt, it was produced at the Cooley Distillery in Riverstown. With a heavily peated style, it was initially unpleasant and challenging to someone more accustomed to the refined sweet peat of Caol Ila – big, heavy, oily, industrial smoke, but with time in the glass it warmed to the atmosphere inside the friendly bar and gave way to a softer smoke, revealing a nice sweetness beneath a layer of peat smoke. Certainly a challenging dram, not an easy drinker, but one worth pursuing and with a reward at the end. The paramedic had a Dead Rabbit. A final dram before the journey home. We left with a bit of friendly advice from the barman: 'Drink Irish, boys'.

As we left Galway, we drove past groups of excited children. They had woken up to the news on morning television that schools and colleges across the Republic would close today from 6pm and there was even talk of St Patrick's Day being cancelled. It was time to leave. Galway City was wiping up a big storm that came careering off the Atlantic, swelling its waterways, and blew us back down the Grand Canal through the peat bogs of the Midlands all the way to Dublin. Across the world, the clouds of a far greater storm were gathering, and nobody had any idea when it would end.

Scotch – Still Going Strong

It's hard not to get away from the fact that Irish whiskey's long, cold winter was in no small part due to a decision not to embrace the art of blending. It's equally true that the survival of the Scotch whisky industry during difficult times, and the spectacular recent growth in sales of single malts, is built upon the unparalleled global success of blended Scotch whisky.

Originating in the early 1800s, the art of blending pioneered by John Walker in his Kilmarnock grocery shop, and Andrew Usher and his son John in Edinburgh, was given a significant boost in 1853 when malts were allowed to be vatted in bond for the first time; this was extended to both malt and grain whisky in 1860. This triggered the meteoric rise of brands such as Johnnie Walker, J&B, Dewar's, Bells, Ballantine's and Cutty Sark, which produced a range of distinctive blends that targeted drinkers and markets across the world. Whisky styles and fashions were rapidly evolving and George Saintsbury, onetime Professor of Rhetoric and

PEAT AND WHISKY

English Literature at Edinburgh University who cared not for the popular blends of his time, identified a change between 1875 and 1920 from darker, sweeter and heavier single malts to something with a lighter and drier taste.[6] Respectable toddy whisky drinkers in the nineteenth century were changing their preference too: 'Traditional toddy whiskies had a very pronounced and oily Islay character that was brought out with the warm water and sugar: the newer blends less so.'[7] With time, cheaper grain whisky, made in the continuous Coffey still, started being added to blends and the malt content gradually decreased from 50 per cent to 20 per cent, more typical of the amount used in a modern blended whisky.

Connections between distilleries making peated whisky and blenders became firmly established in the nineteenth century, and many continue to this day. Famous Grouse, a blended Scotch that is regularly the top-selling whisky in the UK market, has a long relationship with Highland Park. Even during the early days of Johnnie Walker, their blending inventory included an Islay-style whisky, although by the 1860 and 70s, Campbeltown peated whiskies had become a common constituent.

Peated whiskies from Islay are now at the heart of many good blends, but in an essay written in the 1890s entitled *The Art of Blending Scotch Whisky*, Alfred Barnard commented on a nervousness amongst early blenders about the use of Islay whisky, describing it as 'too fat', with 'inexperienced blenders using it too freely and too young'. However, with time its use became 'better understood' and 'more carefully made, the flavour being more delicate and less pungent, owing to the use of dry peats instead of damp ones for drying the malt'.[8] The first written evidence of the company buying Caol Ila appeared in 1897 and the Islay giant quickly became the main peated constituent of Johnnie Walker, especially in blends that had a distinct peaty edge such as Green Label and Double Black. Judicious use of small amounts of heavily flavoured single malt whisky became a mainstay of blended Scotch to the extent that 'A good blend cannot be made without some Islay whisky'. The importance of peat to the makers of good blended whisky is demonstrated by an article in a popular daily newspaper from 1963. Beneath the headline 'The flavour of Johnnie Walker is 3,000 years old', it provides a quick recipe for the formation of peat:

Take a stretch of wild Scottish moorland. Make sure there is a good growth of heather. Leave it to soak for about three thousand years in the damp highland mists; cover occasionally with snow. And what have you got? Peat. Slow-burning, aromatic peat. The peat whose smoke gives so much of the characteristic flavour of malt whisky. And the finest goes into Johnnie Walker.

It goes on to describe to its readers what this strange substance is: 'to the laymen's eye all peat looks like partly melted chocolate'.[9] I have searched in vain for an image of The Striding Man navigating his way across a Scottish peat bog, but as the 3,000-year swirling mist theory for the formation of peat persisted, the marketing team then came up with this piece of whisky foresight:

It's an interesting thought that the heather growing purple out on the moors today will be giving its flavour to Johnnie Walker 3,000 years from now. And that men will still be cutting the peat.[1]

But the rise in the popularity of blended whisky at the expense of single malts was not to everyone's liking and in 1951, Stuart Henderson Hastie wrote that pot-still malt whisky was an 'ideal drink for the man who lives in the hills or for the man that goes fishing or shooting, but for the sedentary office worker of today it is all too heavy'.[10] In essence, he was telling white-collar workers to stick to less full-bodied blends. He would have been even more concerned about the future of single malt Scotch whisky in the late 1950s, when a German company called Rache started making and successfully marketing *rauchzart*. Literally meaning 'tender smoke', this was a trans-European mixture of Scottish malt whisky and 'Germany's tender grains', such as wheat and corn.[11]

The art of blending provided the solid foundations for the long-term survival of the Scotch whisky industry and its future global dominance, and in the context of this story seeded a modern passion for peated whisky, be it blend or single malt. The market share of single malts increased from 0.7% (2.1m litres of pure alcohol or LPA) in 1975 to 10.1% (38.7m LPA) in 2018. This is spectacular growth by any standard, and riding on the back of this growth curve are the peated single malts that in 2018 made up 12 per cent of all single malt sales.[12]

The start of the modern era of Scotch single malts can be traced back to the late 1950s with the relaunch of Glenmorangie and several Speyside whiskies. Then in 1963, the Grant family took the decision to release Glenfiddich single malt into UK and USA markets, effectively opening the door to the single malt category. In London, the Milroy brothers, Jack and Wallace, famously started to sell and really push single malts to the public and trade from their Soho Wine Market in Greek Street as early as the late 1960s. While independent bottlers like Gordon & MacPhail and Cadenheads had been releasing small amounts of peated single malts before the official bottlings, in the 1970s and 1980s Highland Park, Bruichladdich and Laphroaig all had makeovers as single malts.

In 1982, despite the depressed state of the industry as a whole, there were already promising signs for single malts with a 25 per cent increase in UK sales. This was driven by greater customer awareness, marketing, the arrival of new products and the beginnings of whisky tourism. In 1986, The Ascot Malt Cellar was launched by DCL as a marketing initiative with Lagavulin, Talisker, Royal Lochnagar, Linkwood, Rosebank and the vatted Strathconon all under the one umbrella. The concept was to introduce whisky drinkers to regionality and provenance. This was the forerunner to the Classic Malts collection, which DCL, by then in its death throes, launched two years later. Nick Morgan summarised for me the change in direction: 'After the DCL-Guinness merger in 1986, with the business now under new management and new thinking, the company quickly acknowledged the growth potential in single malts. Two years later, they released the six Classic Malts reflecting different regional styles of whisky.' The Islands were represented by a 10-year-old peated single malt distilled at Talisker and the Islay-style by Lagavulin. It is worth remembering that many of Scotland's most famous peated whisky distilleries were closed or mothballed in the 1980s and the whole industry was experiencing a downturn. Survival was the name of the game and this involved an element of risk.

If this was a gamble, it clearly paid off, and what followed in the 1990s was nothing short of an explosion in interest in the single malt category that shows no signs of dimming, Nick Morgan again: 'Peated single malts became the engine room for growth. Demand, particularly from the blenders, regularly exceeded our capacity to produce them – we couldn't make enough of the stuff.' Prior to its closure in 1983 Port Ellen was mainly used by the blenders, but after production stopped, it quickly became one of the most sought-after single malts in the world. Official bottlings of peated single malts became commonplace, but it was only as recently as 2002 that Caol Ila, by some distance the largest distillery on Islay, first released a core range of aged single malt whiskies to the market. Talisker came from almost nowhere to become the most popular peated single malt whisky sold in the UK, suggesting that the launch of the Classic Malts in 1988 was nothing short of a piece of marketing genius. But it is also worth noting that although Caol Ila in terms of its size is in the premier league of Scotland's distilleries, it is very much smaller than Glenlivet. As Scotland's largest distillery, Glenlivet produces 21 million LPA per year – more than the combined output of all of Islay's nine working distilleries.

A comparison of the sales figures between 1982 and 2018 is revealing.[12,13] In 1982, the total number of cases of single malt whisky sold to UK and export markets was around 1.3 million; by 2018, it had jumped to 11.9 million. Led by brands such as Glenfiddich, peated single malt followed suit; sales of Laphroaig jumped more than thirty times, as did Bowmore and Highland Park. In 2018, Talisker had entered the top six single malt brands, selling a staggering 250,000 cases in the UK alone. Remarkably, records show that the distillery that exceeded the sales of all others in the peated single malt category back in 1982 was Springbank, which sold just 16,500 cases.

At Lagavulin, Iain McArthur had a different and interesting take on why single malts took off. 'Money and visitors from abroad,' he told me firmly. 'Nobody had any spare cash in those days, few of us had cars, we could only afford blends. Visitors started coming to Islay, from Germany, Holland, Sweden and particularly the Americans arrived – they were the ones interested in single malts.' But there may have been another less obvious factor. Educating a public keen to acquire knowledge was helped by the publication of books and pocket guides such as *Wallace Milroy's Malt Whisky Almanac* in 1986, which won him the 1987 Glenfiddich Whisky Writer of the Year Award and went on to sell well over 300,000 copies worldwide in several editions.

Many of us are now lucky enough to have more disposable income, so why not indulge ourselves in a way that previous generations could not? Iain's opinion is not one you hear very often and voiced in such a clear way. But it makes sense. Scotch is still going strong and Laphroaig is the most popular peated single malt whisky drunk in the world, well supported by brands such as Talisker, Highland Park, Bowmore, Ardbeg and Lagavulin – all of them originating from the peat-rich islands off the coasts of Scotland. Alongside the well-established names and exploiting the growing passion for smoky whisky are interesting and innovative new ways of working with peated single malt whisky that explore the black box we all know as phenols.

New Age Peat – Sweet and Well-Tempered

Bellshill on the eastern side of Glasgow might sound like the name of a prison, but it is anything but. What was once a landscape of fields and collieries has morphed into the Strathclyde Business Park and at its heart is the operational headquarters of William Grant & Sons Ltd. Six hundred people work here at the hub of the business where whisky is matured, blended, bottled and packaged before being sent to customers all over the world. Tankers and lorries arrive and depart with the names Balvenie and Glenfiddich emblazoned on their sides. It is also the workplace of master blender Brian Kinsman, who met me in the smart wood-panelled reception area displaying some of the company's fine whiskies and brands before taking me upstairs through a lively open-plan office to his large tasting room. Lining the walls from floor to ceiling and covering a work bench at the centre of the room are hundreds, maybe thousands, of labelled sample bottles, and some rather well-known leading brands. Organised in regimented rows, the bottles contain colourless liquids, others lightly tinted with yellows and pinks, and a range of deeply coloured rose, gold and rich earthy browns. Brian is a chemist by training and joined the company in 1997 after deciding that a career in dentistry wasn't his calling. This is his sensory laboratory and the experimentation and inquisitive whisky part of the business. I have come to talk to him about sweet peat and one of his many creations – Ailsa Bay.

The list of names in the company's portfolio – Glenfiddich, Balvenie, Kininvie, Tullamore, Monkey Shoulder, Grant's Family Reserve, Hendrick's Gin – might not speak of peat smoke, but the blend Smokey Monkey does, and so does Ailsa Bay, created at one of the most technologically advanced distilleries in the world, at Girvan on the Ayrshire coast. The marketeers call it 'a modern whisky for the inquisitive', 'inspired by the creativity of science' with 'smoke precisely balanced with sweetness'. I ask Brian to tell me its story:

> The company initially wanted to buy an island distillery to get a smoky whisky for their blends. That didn't happen, so they built a new distillery in 2007 on the site of the Girvan grain plant. Ailsa Bay Distillery has sixteen stills and can produce a range of spirit types: three levels of peated spirit, a light sweet spirit and a heavier sulphur spirit. Peated malt is run for two to three weeks a year. We designed an extremely modern, functional distillery, and the process can be tweaked to alter spirit runs, fermentation times, etc. Still design matches that at The Balvenie in Dufftown. Our aim was to have something that allowed the company to scale up its production to supply its blends.

It took just six months to build. In the natural light of his air-conditioned, temperature-controlled laboratory we taste some of Brian's creations. Two completely clear new-make spirits made from heavily peated malt. Diluted down from the 68% abv, the strength in the spirit receiver, to 28%, the peat smoke really begins to shine as the phenols volatilise in the glass. Cut from different parts of the same spirit run, one of the samples was a softer, less medicinal smoke. A sample from a refill barrel – very light in colour, almost like a young white wine with a tinge of yellow – was sweet and lightly smoky. And then something quite different, a sample from a small first-fill American bourbon Hudson barrel with a much deeper amber colour, richly sweet with a nice balanced level of smokiness. And finally, the Ailsa Bay first release, an example of a Lowland peated malt, bottled in 2015, that combines up to six different cask types before a final period of micro-maturation that brings all the cask influences together.

When Ailsa Bay single malt was first released in 2015, it is fair to say it divided opinion – the geeks were unimpressed with this 'manufactured' expression, but the average whisky drinker warmly approved. On the bottle are the words 'smoke precisely balanced with sweetness' and two pieces of chemical analysis: 21 PPM phenols, and a new measure, 11 SPPM, both analysed in the whisky after maturation. I asked Brian about this new measure.

We have used the SPPM for quite a while as an indicator of sweetness in the whisky. We have selected some of the most influential sweet compounds like vanillin and measure them using HPLC as ppm in the final bottling strength liquid. It is a useful way to quickly plot a flavour map, especially when combined with peat levels. Having said that, the best measure remains nose and taste.

Like any good whisky chemist, Brian disagrees with the way in-grain phenol concentrations are used by the industry since they can decrease by more than 50 per cent during the production process. That said, he tells me that 'phenol concentration changes little during maturation. It's often said that the smoky peated flavours decrease with time in the cask. I don't think they do. What more likely happens is that the influence of the wood increases and gradually becomes the dominant influence.'

The second release of Ailsa Bay Sweet Smoke in 2019 was greeted with wider acclaim; the main difference being that the SPPM levels had been dialled up to 19 SPPM – almost twice the level of sweetness. 'The second release was sweeter because the whisky was older and matured for longer in the cask giving more time for the release of wood sugars, like vanillin,' he said. This whisky also undergoes a final period of micro-maturation in small Hudson bourbon casks.

The creation of Ailsa Bay appears to mark a very clear direction of travel for the industry in terms of science and innovation. I ask Brian where all this is heading. 'The chemistry of wood maturation is a major unknown. Also, better control of the process and improving consistency of the spirit. Saying that, it is always good to have variation. Most of the modern-day master blenders and whisky makers are chemists by training – such an important part of understanding the art of whisky making.' For whatever reason, our conversation now comes full circle to the concept of terroir. To an extent, he buys into the idea: 'Distillery terroir, yes; the place, the plant, the process and the whisky makers all have a major influence on the character of the whisky.' He makes no mention of local barley or local peat. I finished off by asking Brian why he thought the flavours sweet and peat combine so well together in Ailsa Bay and other whiskies?

I think sweet and peat work well together because peat and smoke often have a sweet smell but a dry or bitter flavour. In whisky, the big peated whiskies tend to be very dry on the palate. By increasing the sweetness of flavour through maturation and the use of vanillin-rich casks, we can have a very prominent peaty character on the nose that is complimented and accentuated through a rewardingly sweet taste.

PEAT AND WHISKY - A NEW AGE

The risk of smoke, the reward of sweetness. There can be little debate that Ailsa Bay is a 'whisky constructed by chemistry'; less alchemy and more mixology. So maybe the words of Samuel Bronfman, onetime President of Seagrams, the Canadian distilling empire, 'Distilling is a science and blending is an art', no longer ring true.

In search of more new-age peated whisky, I travelled north to the Isle of Skye, home of one of the most well-known smoky whiskies in the world. This time, I was not heading to Talisker in the volcanic heart of the island but would stay south and drive across the ancient gneisses of the Sleat peninsula to visit the newly born Torabhaig Distillery. I had recently joined the Peat Elite, not a crack team of peat-heads but a members' club that will follow, and most importantly taste, the distillery's journey as its whisky comes of age and progressively evolves.

It was one of those mornings when Skye serves up its own special welcome – leaden grey skies, horizontal rain and buffeting wind. The journey from the car park to the sanctuary and warmth of the distillery buildings was a hazardous one, a frantic dash, narrowly avoiding large puddles on the way. A peaty brown white-water torrent came crashing down from the hills to the sea – the distillery water source. I have come to meet Bruce Perry, global brand manager at Mossburn Distillers, the owners of Torabhaig Distillery and someone who spends much of his life travelling in and out of Heathrow airport. But today, he had landed on Skye.

I wanted to visit and write about a new distillery that is focused on making purely peated whisky, and over a cup of coffee in their warm café, we talk about the journey so far. Spirit only started flowing from the stills in 2017, firmly placing Torabhaig in a group of new-age distillers of peated whisky. What I like about this distillery is their commitment to making the type of peated whisky I adore. They call it 'well-tempered peat' and at its heart are long, slow distillations – three hours fifteen minutes, to be precise – and an unusually high cut point of 64% abv. By cutting high, they reduce the amounts of heavy cresols, and in this way carefully control the key phenol groups in the new-make spirit. 'We have an interest in how the total ppm phenol content is made up, the proportions of constituent phenols, guaiacols, cresols; and how to select the aspects we want in the heart of our distillate.' By 'dialling down the cresols', as Bruce puts it, they are reducing the impact of the medicinal flavours associated with the heaviest Islay whiskies, seeking to create a softer, more rounded flavour profile. There is no fixation with ppm phenols here, just a desire to experiment and explore the subtleties of flavour associated with the different groups of phenolic compounds. The slow distillation used at Torabhaig results in a lighter style of spirit that sounds to me much like the distillation regime applied at Caol Ila, and interestingly the nose and taste of Torabhaig's fledgling whisky has definite parallels to this Islay single malt whisky. Another reason that I like this distillery is that they are using almost exclusively bourbon hogsheads; 80 per cent are first fill, 20 per cent refill. This should work well for the younger whiskies currently being produced by Torabhaig, highlighting that sweet peat combination. There are distinct echoes of Ailsa Bay on the south coast of Skye.

We walk through the rain to the old farmhouse and into the tasting room where Bruce introduces me to an array of bottled single malts and spirit samples. Many of the sample bottles have the names of the distillers on them, 'Kieran's Special', 'Myra's Special', all examples of the Journeyman's Drams, experimental distillations run by members of the team who are given complete control of the whole whisky-making process for four to six weeks. Myra's Special is an unusual unpeated distillation. Most distilleries when switching from peated to unpeated new-make spirit will flush out the system to get rid of the 'sticky' phenolics with a dummy run that is discarded and put to waste. Not at Torabhaig, where the 'waste' distillate with a legacy of peat smoke was casked and is currently being matured for a future release that will be called *Beagan ceò*. This literally translates as 'small smoke' – experimentation is the key at Torabhaig.

There is also a refreshing transparency about how whisky is made at Torabhaig. Bruce told me, 'We have nothing to hide', and that is a huge part of their appeal. This transparency is reflected in the information given on the bottle, with both the maltsters' measure of in-grain phenols and a second value of residual phenols, measured in the new-make spirit or after maturation. The decrease in phenol concentration is significant, often by as much as 75 per cent, caused by their attachment to the grain husk and removal in the draff that often smells and tastes strongly phenolic. A slow distillation regime with high cut points and simple dilution during the whisky-making process also contribute to a drop in the levels of phenols in the final spirit.

I ended the day in the stillroom with Bruce and distiller Kieran Roberts, the architect of Kieran's Special, chewing and smelling malted barley, pouring over sheets of HPLC analyses of new-make spirits from an independent laboratory, and tasting the diluted samples. First up was Myra's Special – unpeated plain malt, soft chewy husks, not a hint of peat smoke and unsurprisingly cereally. Then Torabhaig standard new-make spirit with a starting in-grain phenol content of 75-90ppm and a finishing residual phenol content in the new-make spirit of 38ppm made up of seven individual compounds. It was dominated by phenol, with equal amounts of cresol and guaiacol. Compared to the unpeated malt, the peated barley grains were hard and crunchy – the effect of impregnating the husks with smoky phenolic compounds. The grain had that beautiful smell of sweet peat. Then last but not least, Kieran's Special – a super-peated new-make spirit (138ppm in-grain phenols, 60ppm residual phenols) made from 90 per cent normal malt and 10 per cent black malt, the same barley that is used to make Guinness. The slow smoking process in the kiln produced a very crusty malt that smelt like oily peat ash or even smoked salmon. To Bruce with his well-trained, refined, wine drinker's palate, it smelt like barbecued scallops. Kieran proudly showed me the analysis of his new-make spirit: phenols 23ppm, guaiacols 19ppm, cresols 18ppm.

Like my visit to Brian Kinsman's laboratory at Bellshill, my day spent at Torabhaig Distillery on Skye was an education. Welcoming, patient, open and transparent are all words that quickly come to mind. While both are using analytical chemistry and knowledge of compounds such as phenols to enhance their whisky making, I get a strong sense that there is no desire to anatomise the product with science and step away from those things that make whisky so enjoyable and often surprising. Bruce even corrected me for suggesting that they were using a more science-based approach at Torabhaig, insisting they are making whisky the traditional way. During my visit, I sensed a slight hesitancy or uneasiness to talk about peat. As someone who knew about peat, I was finding I was getting asked more and more on my whisky journey about the subject, about sustainability, management, restoration and the long-term future of this vital whisky-making ingredient.

As I leave Torabhaig, it is still raining heavily, and there is news on the radio of flooding and road closures in the west of Scotland. Grey sky merges with grey sea, torrents of white-water cascade off the hillsides and flow down the roadside. The *dubh* lochans in the peat-rich Lewisian landscape look ominously full. When he came to Skye in the mid-1880s, Alfred Barnard must have encountered similar weather, writing that the rain 'pierces your mackintosh like duck-shot through a boat's sail'.

The Battle of Duich Moss

No shots were fired, no lives were lost but the verbal blows that surrounded the Battle of Duich Moss and its aftermath marked the start of a significant change in the relationship between peat and whisky. The newspapers loved it, with headlines and articles trumpeting the tumultuous Islay summer of 1985, 'Geese Feel the Nip of Whisky',[14] 'Feathers Rustled on Peat Isle'.[15] Duich Moss became a metaphoric battlefield and at its heart was peat cutting and the centuries-old tradition of using Islay peat in the production of smoky whisky.

Facing each other on the battlefront that late July and August were the islanders and DCL on one side, and a group of conservationists led by Friends of the Earth and David Bellamy on the other. The Scottish Office and Brussels became heavily involved and the national press fanned the flames. It was to become the most significant environmental confrontation in the island's history.

The seeds of the conflict were sown in early 1983 when DCL applied for planning permission to extract peat from the south-western corner of Duich Moss. A letter from Dr KG MacKenzie (managing director of SMD) to the Nature Conservancy Council (NCC) in June 1983[16] added significant weight to the distiller's case: 'there is no future in whisky making on Islay if the peat element is inadequate or extraction costs become prohibitive'. In 1984, the Secretary of State for Scotland granted planning permission for peat extraction from Duich Moss, the same year it became designated as a Site of Special Scientific Interest (SSSI). Road construction began in June 1985, with a plan to begin drainage of the area in spring 1986, followed by the commencement of peat extraction two years later.

The special status afforded by the SSSI designation was to protect a population of over-wintering Greenland white-fronted geese; *Anser albifrons flavirostris*, literally the yellow-billed, white foreheaded goose. Duich Moss was the single most important British site with a winter roost of 600 birds. In the eyes of the NCC, Duich Moss was a rare, nationally important peatland occurring in the wetter, western extremities of Europe. It was also *Sphagnum*-rich, treeless, and in the eyes of the distillers, ideal for producing the essential flavours needed to make Islay's famous whisky.

The battle began when the European Commission threatened legal action as an EC directive on the conservation of birds had been breached and the site would now be designated a wetland of international importance. The Scottish Office responded by calling for a suspension of peat harvesting. Prior to a public meeting in late July 1985, Friends of the Earth and TV personality, botanist and naturalist David Bellamy held a demonstration preventing the construction of an access road to the extraction site.

The locals were furious. 'What I resent is interference by people like Friends of the Earth, who are a load of idiots, the Royal Society for the Protection of Birds and the Nature Conservancy Council, who receive £20 million of the taxpayers' money. Now we are being dictated to by Europe,' said Frank Speirs, owner of the Port Askaig Hotel and former councillor who spoke for most of his fellow islanders. A truce in the Battle of Duich Moss was called when the SMD found a new 400ha site, Castlehill Moss, just five miles from the scene of the battle. But after testing, SMD found it was not the right type of peat, went back to Duich Moss and the battle recommenced.

This now became much more than a skirmish and reached a head on 5th August when a well-known English environmentalist, Jonathon Porritt and David Bellamy, were shouted down by angry islanders at a hastily arranged public meeting at Bowmore Hall. It was attended by 650 of Islay's 3,000 population, who endorsed the decision to proceed with peat harvesting on the Duich Moss. The meeting ended in chaos with shouts of 'Go back to England,'

'You are not welcome here – go back to the backyards of England where you belong'. When Bellamy, Porritt *et al.* got back to the Machrie Hotel, their bags were already helpfully placed outside in the courtyard. Their request for police protection fell on deaf ears – they were run off the island.[17]

This was a time of uncertain employment on Islay and nothing was more likely to antagonise the Ìleachs than an Eton- and Oxford-educated son-of-a-lord accompanied by a mouthy TV celebrity/Baptist, who came to the island to preach and threaten their livelihoods. Distilleries were closing or on part-time working as the Scotch whisky industry struggled through a serious downturn in its home and global markets. English interference was also bringing back uncomfortable memoirs of the past – Islay was heavily scarred by the clearances. As a doctoral student at the University of London, Bellamy had botanised on the bogs of Islay and had in a previous life been commissioned by the island's distillers to make a film for them. In later years, he turned full circle and became more extreme in his views, antagonising environmentalists by campaigning against wind farms. Bellamy later became sceptical about the role of humankind in driving climate change and was pronounced a denier.

But the Battle of Duich Moss was over. In 1986, the NCC presented a nine-part dossier to the Secretary of State for Scotland on the importance of the moss for nature conservation highlighting the damage that would be caused by drainage and peat extraction.[16] It is direct and to the point, calling moss water the 'life blood of the bog, with the wettest lower catotelm[18] likened to our body's internal organs that must never dry out. The upper acrotelm[18] is the protective skin, maintaining and shielding the body's own moisture.' The European Commission was left in no doubt about the fragility of these bogs with the catotelm in a suspended state of fluidity that can occasionally break out as 'bog bursts'. Further scientific surveys and research followed aimed at finding a suitable alternative site. In 1993, after several years of discussion, Duich Moss, or *Eilean na Muice Duibhe*, formally became a National Nature Reserve, designated a wetland of international importance under the Ramsar Convention. The moss was saved and its habitat, birds and plants protected. The temperature in Islay had dropped and peat extraction for Port Ellen Maltings switched to Castlehill Moss. Maybe there was an inevitability about the choice as SMD had owned the peat extraction rights at Castlehill prior to the 1980s, which they exchanged for those on Duich Moss.

The morning Paddy and I visited the Duich Moss National Nature Reserve, all was quiet. It was big, open and windy. Lying just 30m above current sea level, it is underlain by an old raised beach that slopes imperceptibly down towards the coastal plain and Loch Indaal. Geese were about, but not the protected kind. I found no signs proclaiming its protected status and apart from the cars surfing across the rises and falls of the pot-holed, patched up peat road, I did not meet a soul. A bright orange Hebridean Air Services twin-engined plane climbed noisily into the sky heading north, then east, back to its base at Oban on the mainland, after arriving a few minutes earlier from the tiny island of Colonsay.

We tried to penetrate an old area of peat cutting along a disused track built high and dry above the moss – it was now overgrown with willow and birch and when we drifted off-piste, we rapidly became sunken and wet-footed. A good sign: the bog was wetting up well despite there being no attempt to restore the site. Small *dubh* lochans were reforming, *Sphagnum* regrowing, accompanied by grasses, shrubs and the ubiquitous, ever-present bog cotton. Sundews and the early purple orchid (*Orchis mascula*) added splashes of exotic colour to the moorland community of plants. The layout of the old hand-cut extraction cells was still evident – a succession of parallel deep drains, overgrown peat banks and dry ridges, repeated time and time again, and the occasional rusting piece of machinery. Almost forty years after the whisky industry and environmentalists went to battle, Duich Moss now had an impenetrable feel about it.

Global Warming Becomes Global Heating

Apart from those who utilised it purely as a resource to be harvested and burnt, for a long time peat was a subject few people cared about or even noticed. Bogs were regarded as barren land that could be cured by drainage and the addition of lime, made productive and used in some purposeful way. Then something changed. People started to have more leisure time, opened their eyes to the great outdoors and began to appreciate and know more about what lay beneath their feet. Global warming arrived and we all began to learn and care about the role of peatlands in mitigating rising atmospheric carbon dioxide concentrations. People started to visit them. Quaking bogs and sodden mires became peatland ecosystems of significant environmental importance, beauty and home to rare species of animals, birds, plants and insects. As they started to be researched, surveyed and measured at a local and national level, they became better understood, designated, managed and protected by law. They are nature's own carbon capture and storage system, but only a healthy, functioning bog will be effective at drawing down carbon dioxide from the atmosphere. As the world begins to appreciate the important role peatlands play in mitigating against the effects of climate change, there is a growing respect that didn't exist at the end of the twentieth century for this former fossil fuel. But it is more than a change – it is a seismic shift.

Nowhere demonstrates just how clearly our recent perceptions about peat have changed more than the island of Ireland, and on our journey across the country, it was striking to see for ourselves a remarkable transition taking place. Like the people of Islay, the

Irish Turf Cutters Association has had recent skirmishes with the European Commission, but power from peat was a project on a previously unthinkable scale. Portarlington power station was the starting point of an economic programme that by 1977 saw Ireland generate a third of its electricity from the burning of milled or sod peat. The power station no longer exists. It opened in 1950 and during its forty-seven-year lifetime consumed 195,000 tonnes of dry peat each year, enough time to form five centimetres of peat.

'The bogs…are the true gold-mines of Ireland,'[19] wrote Kerr in 1905, and in the 1930s after the formation of the Irish Free State, the new republic's founding fathers had a vision to turn peat into energy by draining bogs, strip-mining the dry peat and then burning it. They created the Turf Development Board with a mission to exploit the country's bogs and reduce the nation's dependence on imported fossil fuels, as well as generate jobs in poorer parts of rural Ireland. Following coal shortages in World War II, Ireland's *Bord na Móna* or Peat Board was created, and what followed was a countrywide re-engineering of Ireland's bogs accompanied by the construction of power stations with a supporting infrastructure of narrow-gauge peat railways to link the two together. Bogs have always been an important part of the Irish psyche, although they were traditionally seen as large inaccessible wastelands, a refuge for bad people, where bad things happened. As far back as the late seventeenth century, there was a desire to drain them, turn these barren lands into something useful, cultivate and farm, and maybe even remove them from the landscape altogether.[20]

We drove past Edenderry with its tall red-and-white striped chimney, a power station in transition from burning peat to generating 100 per cent of its electricity from renewables. Further west, the bogs had been transformed into opencast peat mines of a size that I have never seen before. They were feeding the West Offaly power station, the largest peat-fired plant ever built in Ireland, peacefully smoking away when we passed by on a Sunday afternoon. Railway tracks appeared from nowhere carrying hundreds of peat-filled trucks, their contents waiting to be disgorged into the power station's furnaces. In 2018, *Bord na Móna* announced that they would end their use of peat to generate power and close 'active' bogs.[21] This has turned into an 'accelerated exit from peat', as the state-owned company speeds up its journey from fossil fuels to renewable energy. Wind is the future and across the flatlands of Ireland, groups of powerful, large turbines now harvest the energy that once came from the soil of 'Old Ireland'.[19] West Offaly power station wasn't commissioned until 2005 and closed in 2020, a short working life that marked the end of a great national project that had lasted seventy years.

Peat has entered a new age and so has peated whisky. Our climate is changing, humankind will need to adapt and, like it has done in the past, the whisky industry will adapt too. The science that underpins our understanding of climate change is well grounded and goes back to the early 1800s, when researchers realised that the Earth's atmosphere was warmer than it should be. Its constituent gases were trapping incoming solar radiation and in 1896, using the principles of physical chemistry, the Swedish scientist Svante Arrhenius became the first person to calculate that increased human emissions of carbon dioxide would lead to a rise in the surface temperature of the Earth. While Arrhenius established the scientific basis for our modern-day understanding of the greenhouse effect, the evidence linking the two came much later.

Two years after I was born, in March 1958, Charles David Keeling started measuring atmospheric carbon dioxide concentrations at the observatory on the slopes of the Mauna Loa volcano in Hawaii. At the time, they were 313ppm and these became the first points on a trend line that we now know as the Keeling Curve – the single most important graph ever produced by a climate scientist.[22] When I started my PhD in 1979, the concentration had increased to 334ppm. As I write these words more than forty years later, it has reached 415ppm and will continue to rise further.[23] According to the National Aeronautics and Space Administration (NASA), average global temperatures have risen by 0.83°C over this time – in the polar regions, warming has been much faster. Should we be worried? Absolutely. Scientists like myself are typically cautious creatures who only make firm statements or reach a conclusion when the evidence is overwhelming. Well, the evidence for human-induced climate change is overwhelming and I have not met a creditable scientist who thinks otherwise. We have not only created a hotter world, but the extremes of climate, be they drought or flood, often now seen on a biblical scale, are becoming more common. The once-in-a-thousand-year meteorological or hydrological event is now occurring on decadal or shorter timescales. Period. Global warming has become global heating.

Professor Roxane Andersen is one of my favourite people for obvious reasons – she is passionate about peat and quite fond of whisky. A native of Quebec City, who, after visiting various European countries, decided to make Scotland her home and its peatlands her cause. A biochemist and ecologist, she cut her teeth on the *Tourbières du Québec*, drank whisky with her supervisor in a field hut in the middle of a bog and quickly 'decided I quite liked it'. She came north to work in the peatlands of Caithness in 2012 and set up the Flow Country Research Hub that attracts scientists from around the world to this remote corner of north-east Scotland. Roxane lives next to a peat bog, has a garden full of midges, married Gearóid who hails from the bogs of Ireland, and has a young daughter who knows the names of more *Sphagnum* species than I do. I asked her how she excites children about peatlands. 'It's easy,' she replied, 'I show them that the more you look, the more you see.' It's a sentence that encapsulates beauty, complexity, scale and wonder.

PEAT AND WHISKY – A NEW AGE

Roxane's team of scientists and students, along with her network of collaborators, have built something different and special in the Flow Country. They have applied peatland science to underpin considerable restoration success through engagement with landowners and the wider community. By promoting the benefits of restoration, they have brought onboard those who in the past would have been described as sceptics. With the passion of a Québécois, she tells me, 'It has become a model for the rest of the world, Scotland is leading the way', and frequently takes calls and hosts visits from peatland practitioners from around the world. Climate change is a topic never far away in our conversation and long gone are the days when peatland biodiversity was the only topic in town. 'Peatlands obviously still have an intrinsic value, but climate change is the single issue that has brought peatlands to the wider attention of the public and politicians – peatland restoration is one of the tools at our disposal to fight global warming.' People are thinking big in this part of the world and plans are afoot for the Flow Country to become a World Heritage Site – watch this space.

The whisky industry in Scotland is in a fortunate position because the country is already a global leader in peatland conservation, with research in restoration feeding into government policy with a strong buy-in from the landowning community. In short, the Scotch whisky industry can tune in to the existing knowledge base. But I also wanted to get a global perspective on how recent perceptions about peat have changed and gauge how the science had evolved, so in 2021, I caught up with one of the most widely respected peatland scientists in the world. Professor Nigel Roulet, based at McGill University in Montreal, is someone I have known for a long time. He served on Intergovernmental Panel on Climate Change (IPCC) from 1995–2009, a period that culminated in the ground-breaking AR4 Report in 2007, and shared the award of the Nobel Prize with his fellow scientists. Like many Canadians, he is a great lover of Scotch single malt whisky, especially the peaty stuff, and is one of thousands of fans of Laphroaig who own a very small piece of an Islay bog. I started by asking him whether it was still correct to refer to peat as a fossil fuel.

> The perception that peat is a fossil fuel has changed since the time when it was burnt for energy production in large power plants in countries like Ireland, Finland, Sweden and Russia. For 'economic convenience', it was once classified within a group of fuels including coal and oil that climatologically act in the same way. The closure of almost all peat-fired power stations now means that peat is no longer considered a fossil fuel in an economic sense, but more as an organic soil. Peat is not a geologic deposit like coal and oil that are both fossil fuels with a non-renewable, finite and fixed base. They take tens or 100s of millions of years to form – peat formation can now be measured on yearly, decadal or millennial timescales.

In short, once coal and oil are extracted from below ground and used, there is no way back; this is not the case for peat. I then ask Nigel whether in the light of what we know about peat accumulation rates, can it now be considered a renewable resource?

> We have moved away from the idea that peat is a non-renewable resource. In the early days, climate policy was based on a 'computer-limited timescale' focusing on a horizon of 100 years. The simple reason for this was that computers in the early 1990s were not powerful enough to process the amounts of data needed to look further into the future. Thirty years later, modern super-computers can comfortably do the maths to predict scenarios in the coming 250–300 years. These longer climate trajectories now allow the models

to include materials like peat that renew on relatively short timescales. The second thing that changed was the science. At one time, I thought it was almost ecologically impossible to restore peatlands to recreate an ecosystem that once again acted as an effective carbon sink. I was wrong, and with the benefit of measurements made by research scientists in Canada and Europe, we now know that it is entirely feasible to restore peatlands in timescales as short as ten to twenty years. Peat bogs are a lot more resilient than we first thought and act as self-regulating systems – in wet years, they respond by taking up more carbon; in dry years, they go into standby mode until conditions become wetter and more favourable for peat growth. If a peat bog is completely harvested down to its geologic base, it will not recover.

That is good news for governments like the one in Scotland who are spending millions on restoration efforts. But it has been clear to me for some time that climate change driven by global warming is a far greater threat to the survival of global peatlands than peat extraction, so our conversation moves on to the UK and our changing climate, which is seeing more extreme weather events,[24] be they floods, droughts, or heatwaves:

Although you don't have melting permafrost or disappearing sea ice like we have in the Canadian Arctic, some of the greatest changes in global climate are happening right now in the UK. The climate envelope for peat formation is smaller for the UK, which means it is very sensitive to yearly changes in precipitation. Unlike large continental land masses such as Canada, the UK has a wet, maritime climate. This means that peat formation is dependent on maintaining a sensitive balance between precipitation and evapotranspiration. A shift to a more continental climate with more easterly winds is likely to lead to drier peatlands and an increasing frequency of drought and fire.

The environmental and green credentials of the whisky industry are impressive and have been for some time. This inbuilt resilience is partly due to its long history, but most importantly, to produce whisky you have to understand the natural cycles of the growing seasons and water availability. For this reason, it is better placed to meet the significant environmental challenges of the future associated with resource availability and global climate change. The challenges are numerous and go far beyond the issues that surround peat extraction and its use by the whisky industry. From talking to maltsters, producers of distillery peat and the distillers, I estimate that current peat use by the Scotch whisky industry is around 6,000 to 7,000 tonnes of dry peat a year, less than 3 per cent of the total peat use in the UK. Most of the rest is earmarked for the horticultural sector, with little now burnt for domestic heating. The International Union for the Conservation of Nature UK Peatland Programme give a higher value of less than 4 per cent; the official SWA number is less than 1 per cent.[25] In 2023 the SWA published its long-awaited 'Commitment to Responsible Peat Use' that aims to chart a clear path for the sustainable use of peat for the industry. [26]

In this context, it is also important to recognise the significant historical changes that have taken place in peat use by the whisky industry in Scotland. In the late 1880s, there were around sixty-five to seventy working peat bogs being used, a number that grew further at the turn of the century with the building of new distilleries. When floor maltings closed in the 1960s, many distilleries no longer needed their local bogs and they fell into disuse. At the last count, there are just seven to eight left in Scotland that provide peat for the whisky industry.

Peat use by the whisky industry has clearly changed since the time it was used both as a fuel to fire the stills and dry the malted barley. We can only guess at the numbers, but the amounts 150 years ago would have been huge. For example, in 1885 Bunnahabhain distillery cut 2,600 tonnes in a single year. In the early part of the twentieth century, company records show that individual distilleries across Scotland were using between fifty and 1,100 tonnes of dry peat a year. This equates to a total in the region of 31,000 tonnes across the industry as a whole, five or six times higher than current levels of peat usage. But will this downward trajectory continue, particularly in the face of a rise in popularity of peated whisky? With a ban on the sale of horticultural peat in garden centres likely in 2024, peat use by the whisky industry will move firmly into the limelight.

Drainage of a peatland has been likened to a knife wound – the bog will only survive and begin to heal if the wound is plugged and the life blood of the bog retained. The fact that up to 80 per cent of UK peatlands were damaged or 'wasted' when surveyed in 2015 suggests that there is significant national capacity to enhance carbon storage.[27] Peatlands need to be looked after and nurtured, not neglected. Responsible and sustainable peat use must be the goal for the Scotch whisky industry, not least in terms of the benefits that it will bring to water quality and supply security. The solution for the whisky industry is to invest in a programme of peatland restoration that captures, on an annual basis, at least the same amount, but ideally more, carbon from the atmosphere than extraction and combustion releases. Recent signs are encouraging with Beam Suntory's Peatland Water Sanctuary a welcome initiative that aims to restore 1,300ha of degraded Scottish peatland, and by doing so sequester carbon that offsets the amount of peat it uses to make its peated whisky.

My hope is that the industry will avoid greenwashing the peat issue; planting trees to offset fossil fuel emissions will not restore Scotland's damaged peat bogs. To put this into some kind of context, the peatlands of the Flow Country in northern Scotland lock up more than double the amount of carbon stored in all UK forests. These are the places where the whisky industry can intervene and make a real difference, with a mosaic of peatland restoration sites of different ages, that will deliver a sustainable future for peat. The prize is enormous.

No More Peat?

In 2015, the whisky world was temporarily reeling from the headline 'No More Peaty Whisky From Islay'.[28] It was the 1st of April, a joke, but what once seemed implausible or unthinkable has become, whisper it, possible.[29,30] Six years after I last walked past its entrance on my journey across the Flow Country, I returned to Berry Croft and was greeted by a large sign with the words 'No More Peat'. Such is the dexterity of people in this part of the world that it seemed a successful peat-harvesting operation had turned into a strawberry farm and jam-making business. I was left in no doubt when I was greeted by a happy young toddler dressed in a white coat decorated in bright red strawberries and a matching pair of strawberry wellies. The girl's grandma, Tricia, the maker of all that jam, told me their days of cutting peat on Causeymire were nearly done. For many years, the Sutherland family had been supplying Caithness peat to Bairds maltings in Inverness to make Bruichladdich's superpeated Octomore.

I drove out to Causeymire to see what remained of a once sizeable peat-harvesting operation, the last throes of industrial peat extraction in Caithness. In the past, it was sold to distilleries, used as a domestic fuel in Newcastle and London,[31] and in the

1980s shipped to Sweden and the Faroe Islands.[32] When I arrived, I found a collection of rusting machinery – graders, extruders, conveyor belts – now more junkyard than machine park. A final pile of compressed peat sausages is waiting for a lorry to take them south to provide the phenolic power for this year's batch of Octomore. Causeymire is now home to more than fifty fast-spinning wind turbines – there were none when I first came here more than twenty years ago. The meaning of peat power in Caithness has been redefined.

It is not hard to find evidence of the difficulties that might lie ahead for the Scotch whisky industry. Lying 50m above the waters of Scapa Flow is a field of freshly cut peat lying on a weathered, wet and puddled platform of pure white Orcadian sand, cut away from its geologic base. The turfs are neatly laid out, some black and amorphous, others light brown and fibrous. There are no signs of trees and no remains of ancient wood. The older cut peats have cracked in the dry Orcadian air, their surface like the bark of an ancient conifer or the skin of a giant lizard.

At the edge of the field of peats, I come across a rather half-hearted attempt to restore the bog. Several large corrugated plastic sheets, metal rods and wooden posts driven deep into the ground are attempting to dam, slow the water movement and wet the hill. Encouragingly, Highland Park and the RSPB are trialling several restoration methods, but rather than a plausible route to salvation, this one is little more than an eyesore on the landscape. What I see does not fill me with hope. Dry hill slopes are very difficult places to regrow peat, and when it is cut down to its base it is practically impossible.

Early on that morning, I had met the contractor and his excavator who 'hand cut' the peats, his work done for the year. The machine, equipped with a special cutter, had won all the peats for the season in a matter of days. In the 1800s, it took ten men, supported by women and children, weeks of hard labour to hand cut peats on Hobbister Hill for the distillery. The old peat banks are still evident, covered by a thick blanket of heather. Peat cutting has clearly exacted its toll on the hill and the contractor had told me that it was becoming more difficult to find good peat. It was getting a bit thin and the 'good uns', as he called them, were gone.

The distillery is 11km (7 miles) away on the outskirts of Kirkwall and when I arrived, the two kilns were fired up, one smokeless and burning coke, the other belching aromatic peat reek that drifted across the nearby fields and houses. Diluted in the fresh air and mixed with the scents of wildflowers, it is sweet and aromatic, everything Highland Park should be. I had visited the distillery on a previous occasion more than thirty years ago, but nothing had changed. It is the definition of consistency and tradition – floor-malted barley, peat and coke-fired kilns and working pagodas, still used for the purpose they were designed for.

One hundred years ago, it was not uncommon for distillery owners to boast about their 'inexhaustible source of excellent peats' or 'fine peat mosses', but things are changing. In researching this book, I have been asked by whisky people if we will ever run out of peat. Around 20 per cent of the land surface of Scotland is covered by peat, so the answer is unequivocally no. I have also been asked about the sustainability of peat use by the industry, which suggests we are beginning to see clear signs of a shift in the tectonic plates. If there is 'no more peat' to harvest, what are the alternatives? What if the whisky industry is severely rationed in its use of peat, or worst still, prohibited from using it?

In 1798, Adam Whyte, an excise officer in Edinburgh, gave recorded testimony to a committee charged with reporting on the duties levied on distillers in Scotland.[33] He was asked: 'Is there or was there at Cannon Mills any Still or Apparatus for making Whifkey with the Flavour or Smoke of Peat?'

He answered in the affirmative. 'There certainly is an Apparatus which I have frequently feen at work, for the Purpofe of extraction the Effence of Peat by condenfing the Smoke arifing from it, which at pleafure they mix up with Quantities of the Spirits of the Wafh to give them the Peat Flavour. Mr Henderfon informed me it was mixed with the Spirits. I never faw it mixed, but have tafted the Spirits after they had acquired the Peat Flavour.' More than 200 years ago, the Cannon Mills Distillery in Edinburgh was manufacturing a concentrated peat smoke essence to flavour its whisky.

Created in 1895 by Ernest Wright, a Missouri pharmacist, liquid smoke was made by condensing wood smoke in contact with cold air and collecting the droplets to cure meat. As an alternative to direct smoking, it was first used as a preservative, but by the 1970s, it had become a common flavouring for cookouts and barbecues in the USA. A similar process can be used with peat instead of wood to distil off a peat smoke concentrate, which is then sold as liquid peat essence. Peat smoke can also be extracted directly into a liquid and then distilled to concentrate the peat smoke in liquid form. Modern whisky makers like Lark Distillery in Tasmania have experimented with a post-malt peat-smoking process involving trapping peat smoke in balls of cotton wool and placing a group of infused smokeballs into the wash still. This gets about half the levels of phenol into the spirit compared to the traditional method and takes a little longer.

With our current knowledge of the chemistry of the key flavour compounds in peated whisky, it is theoretically possible to synthesise peat smoke flavours by mixing known amounts of individual, naturally occurring phenolic compounds, readily extracted and widely available from industrial suppliers. Current SWA rules don't allow the addition of flavourings to whisky spirit, apart from small amounts of caramel that can be added as a colouring. Wood smoke is being increasingly used in countries like Sweden, USA and Australia to add flavour to whisky, but not in Scotland, where there appears to be a legal barrier to its use. So thankfully, and if the rule makers allow, there are alternatives to using peat for the Scotch whisky industry, but no one realistically wants to go down that path and end the centuries-old link between peat as we all know it and whisky.

Peated whisky has enjoyed two golden eras, the first rooted in the illicit pot stills of the Scottish Highlands and Islands, when peat was the only source of fuel available to the distillers. This produced a heavy, smoky whisky that was highly sought-after and regarded as a premium product. The invention of the continuous still and the arrival of blends in a growing global market, particularly in North America, resulted in a shift to a lighter, more widely acceptable and palatable style of whisky. While this marked a temporary dimming of the peat flame, it was never extinguished, and it could be argued that the wider consumption of blends with a touch of peat smoke sowed the seeds for what followed. At the beginning of the twenty-first century, we are experiencing a second wave of peat-fuelled whisky consumption, with peated whiskies appearing across the world and attracting a new and younger breed of whisky connoisseur to the fold. Building on its roots in Scotland, it is currently produced across the continents of Europe, Asia, South and North America, Australia and Africa. Although the Antarctic still remains to be conquered, in 2009 peaty Ardbeg became the first whisky in space when it journeyed around the earth for three years onboard the International Space Station.

In the modern world of Scotch whisky, traditional boundaries between the established five whisky regions have become blurred and some would argue no longer relevant. It is now difficult to find a distillery that doesn't have some sort of peated expression in its core range. Some lament this change, seeing it as a dilution of the strong peaty character that at one time defined whisky regionality in Scotland; others see it positively, as evidence of the growing appeal and wider acceptance of the smoky flavour profile by the whisky drinker.

In 2009, the whisky writer Ian Buxton wrote, 'The peaty taste of mainstream whisky, especially blends, is for the most part a thing of the past'.[34] So is peated, especially heavily peated, whisky just a phase we are going through, a fashion? I think not. Peated whisky can't be dialled up like this year's must-have pair of trainers or fashion accessory. Aged peated single malt whiskies have to be laid down years in advance of coming to market, so the industry has either predicted this or subtly marketed and planned for a change in consumer taste by attracting a new group of consumers to the fold. While it is likely to be a combination of both, it is also safe to say that a significant amount of market research lies beneath the surface. There are all sorts of reasons why peated whisky is here to stay and will continue to grow in popularity across the world, but more than any other of the styles of whisky, peat smoke is a natural, strong flavour that conjures up something rather different. The secret of whisky as a brand is its diversity, and peated whisky is part of that success.

PEAT AND WHISKY

So why do so many of us love the smell of peat smoke and the aromas and tastes of peated whisky? Is it a terroir thing – something deep and earthy that links us to a place we want to be or have fond memories of? The eminent whisky scribe Charles MacLean muses how 'perhaps the big Islays, the smokiest of all malt whiskies, recollect the whisky of the past', they have 'authenticity' and 'heritage'.[35] Few of us have tasted the whiskies of the past, so I asked Dave Broom the same question and he had a different response:

I think that the smell of smoke promotes an atavistic response in us all. We're hard-wired to react. There is a fear factor, of course, but also a feeling of pleasure, which could be due to smoke being used as a food preservative for millennia. It offers comfort. Neil M. Gunn wrote that, for him, the smell of wood smoke was danger – moor burn, wildfire, etc., whereas the smell of peat was the smell of home. I've had the same thing said to me by people who were brought up with peat fires burning in the hearth.

As well as this unconscious response to the aroma, I also think that a peaty whisky gives a drinker – a new one, especially – a feeling of confidence. People are fearful of speaking about whisky because they feel they are unable to articulate what they are smelling. I think it simply comes down to them worrying that they will say 'the wrong thing'. Peat is easily identifiable, it allows a new drinker to feel they have a grasp on this new world. Being able to name the aroma gives them that confidence boost. They aren't stupid, they can smell, they do understand. The same happened with wine when it was democratised in the 1990s. Big flavours arrived that gave drinkers something to latch on to: Cabernet – blackcurrant, Chardonnay – vanilla and butter.

There was also the vindaloo effect – some people will always want to try the most extreme flavour going. It was fascinating to see what happened when United Distillers launched its Classic Malts in the 1980s. Logic suggested that new drinkers would start at the light unpeated end, with only a few ending up at Lagavulin. The opposite happened – the 'difficult' flavours of Lagavulin, and to some extent Talisker, became the most popular.

This is what makes peated whisky so fascinating. In our clean, modern, sanitised, smoke-free lifestyles, do we still retain a distant ancestral memory of smoke from a time when we cooked with a peat or wood fire, warmed ourselves in its heat and cured and preserved our food in its smoke? Maybe somewhere in the deep recesses of human memory, we still yearn and search for those lost sensory signals and find comfort in a whisky infused with the smells and flavours of peat smoke. But it's okay not to overthink this and just reward yourself with the 'incense of slumbering ages'.

Postscript

'On the windy edge of nothing.'

My hope is that the relationship between peat and whisky will never end and what we have witnessed so far is just the first part of a story that will naturally change in the future as it has done in the past. The American anthropologist Stuart McLean, writing about Europe's 'muddy margins', refers to peat as 'black primordial goo'.[1] He could not be more wrong. Peat is a remarkable material – a fragile life-giving source of heat and water, a substrate upon which whole ecosystems depend, a sponge for atmospheric carbon dioxide, a precious living anthropogenic record and a medicinal balm. In the past, peat was used as a fuel by the whisky industry; now it is used exclusively as a flavouring. If you are a whisky drinker, and you should be one, peat is an irreplaceable source of aroma and flavour – an essential essence that reaches across the world from its beginnings in Scotland and Ireland to Australia and India. So where to end this story of peat and whisky?

We had just dropped down through the clouds and after a steep left turn Sofia spots the runway and the Atlantic Airways jet lands on a plateau of black basalt lava flows on Vágar, one of the eighteen islands that make up the archipelago of the Faroe Islands. This is the location the British military chose to build a 'concealed' airstrip after they occupied the islands in response to the German invasion of Denmark in 1940. They also brought to the Faroese people a love of many things British including Cadbury's chocolate, Tunnock's Caramel Wafers and, you guessed it…fish'n'chips. Rather confusingly for the Faroese, the visitors also insisted on driving on the left in Vágar. Located 320km (200 miles) due north of the Butt of Lewis lighthouse, we have come to visit the Atlantic's newest whisky island, or 'The Island of Sheep' as John Buchan called his 1936 novel.[2,3] The Scottish writer and poet Eric Linklater called it 'the Windy Edge of Nothing'.[4]

For very good reason, the Faroe Islands, with their horizontal black stratified cliffs and mountains, bear an uncanny resemblance to the better known whisky-producing islands in the North Atlantic. Geologically connected to the young volcanic rocks of western Britain including the remote islands of Rockall and St Kilda, they formed from the same mantle hot spot that led to the opening of the North Atlantic Ocean some 60 million years ago. The flood basalts of County Antrim, Skye and Mull that poured out of fissures and volcanoes are all rooted in the same geological upheaval that gave birth to the Faroe Islands and continues to this day in neighbouring Iceland. The early lava fields of the Faroes would have at one time covered a much larger area than they do today, but millions of years of relentless pounding by the North Atlantic have created a fragmented landscape of individual islands. More recently, active carving glaciers followed by a rise in sea level at the end of the last Ice Age created a landscape of deep flooded fjords and dramatic steep-sided basalt cliffs and headlands. Severed by time, those ancient geological connections have now been re-established by fast modern ferries, suspension bridges and deep-sea tunnels, which connect the islands and have helped build a cohesive, progressive twenty-first-century country.

As a geologist who had walked and surveyed the volcanic islands of the west coast of Scotland, Sir Archibald Geikie was attracted to the basalt lava flows of the Faroe Islands and was struck by their similarities to the cliffs of Skye and Mull. In his *Reminiscences* published in 1904, he describes possibly the first time that Talisker whisky was tasted by the Faroese.[5] After navigating through dense summer sea fog, his steam yacht *Aster* made landfall in 1894 on the southernmost island of Suðuroy. Geikie and his crew were immediately examined by a local doctor before being allowed to disembark; the islanders feared smallpox, which at the time was endemic in Scotland. After pronouncing a clean bill of health, the doctor was offered refreshment and he chose whisky, presumably for medicinal reasons. After 'tossing it off as if it had been so much as water', he became temporarily speechless and gasped for breath. In Geikie's words, 'If he had never tasted Talisker whisky before, we believed he would not forget his first experience of it'. Talisker, the peppery 'bonfire on the beach', has now gone on to become one of the best-known brands of peated whisky in the world with a total of thirteen different expressions on the global market. Maybe it all started in the fog of Suðuroy in 1894; author Derek Cooper called Talisker 'not a drink, it is an interior explosion'.[6]

Peat and whisky have now entered a new age where whisky of all types can be made just about anywhere. Distilleries no longer need local fuel sources to power their stills and raw materials – yeast, peat, malt and peated malt – can be shipped across the world. The Faroe Islands have in abundance two of the most important whisky-making ingredients. First, clean surface water and lots of it, which comes cascading vertically down the sides of the basalt cliffs in rivulets as soon as it starts to rain. Cereals aren't grown today, but the early Norse settlers grew barley, using a strip cultivation method unique to the Faroe Islands called *reinavelta*. And second, the Faroe Islands have peat and no less than twelve different types to choose from. This is one of the reasons the islands have a strange, almost mythical attraction to peatland scientists, the greatest of all being Petur Jacob Sigvardsen, who in 2006 published an epic 1,992-page thesis entitled *Torvið í Føroyum – í søgu og siðsøgu*, (*Peat in the Faroe Islands in History and Cultural History*). It ran into five bound volumes.[7]

For over 1,000 years in a treeless country, peat was power in the Faroes, but not everyone had access to it, and in an attempt to create equality there were acts of parliament, laws, taxes, and peat politics were to the fore. From the peatfields of the deep valleys and heads of fjords came the fuel for the *roykstova* in homes throughout the islands. Turfs were transported in creels by Faroese pony, on cableways strung from the high basalt cliffs to the shores of the fjords below, and then by boat, but only on a 'peat-calm sea' as the local saying goes. In many places, the peat was literally 'scalped' or 'flayed' from the land leaving only bare rock or mineral soil behind. The peat-cutting season was short and it was said in the Faroes that the arrival of the tern and the peat spade go hand in hand, around the feast of St Halvard on 15th May. The workers in the peatfields were fuelled by rye or unleavened bread, dried whale meat, dried fish with potatoes and blubber, fish balls with lamb's tallow (rendered fat), meat cakes, puffin, guillemot, all washed down with a bottle of milk. The peat cutting and drying season, which ended in late July, is now a thing of the past. With the mining of Faroese coal from Suðuroy supplemented by UK coal including shipments from the Brora Coalfield,[8] peat use declined throughout the twentieth century, and by the 1950s had been consigned to history.

We had arrived in the rain on the day before midsummer. Our taxi driver had told us 'We are basically all the same, the Faroese, the Scots and the Irish'. I asked him, what was the best month to visit the islands? After a thoughtful moment, he replied, 'Well, the best month is February.'

'Really, February, not June?' I said, somewhat surprised.

'Yes, February. It only rains for twenty-eight days.'

Our welcome pack from our host Litzi included two Tunnock's caramel wafers. We felt at home.

On Midsummer's Day, we rose early to catch the morning bus and head north out of the capital, Tórshavn. One bridge, three mountain tunnels, one deep-sea tunnel and ninety minutes later, we arrive in a 'blizzard of rain' in Klaksvík on the island of Borðoy in the northern fjords. It was living up to its reputation as the wettest town in the Faroes, and with a temperature of 8ºC and a windy, wet fog blowing down the fjord out of the north, it felt to me more like the High Arctic than the Subarctic.

Klaksvík is home to *Føroya Bjór* a family-owned brewery founded in 1888 making beer from 'Faroese mountain water, Danish malted barley, hops and Caribbean cane sugar'. For the first time, in 2011 it became legal in the Faroe Islands to make alcoholic drinks with a strength greater than 5.8% abv, and like many breweries of the past, owner Einar Waag started distilling spirits. First came gin, vodka, the Faroese party drink *Akvavitt*, and then Einar's Cask *Akvavitt*, both traditionally drunk from a ram's horn. Now he is distilling whisky using malted barley shipped into Klaksvík harbour from Simpson's maltsters in the UK. Einar is making a light, fruity spirit with unpeated malt with long, four- to five-day fermentations. With a nod to the future ('the world has gone crazy for peated whisky') and maybe to the peat-rich landscape of his home, Einar is also using a small amount of peated malt.

The German-made pot and columns stills dominate the entrance to the brewery and look out onto the inner harbour. We visit an old waterside boathouse, now a cask store. Amongst a collection of small barrels and quarter casks, mostly sherry with some bourbon and French wine, is a cask of *Whisky Roykt*. I ask Einar if there is a history of whisky making, legal or illegal, in the Faroe Islands. 'No, not at all. There was a Dane who came to the islands in the 1950s and started to make moonshine. The customs knew what he was up to but couldn't find anything. Eventually they located his still, underground, in a septic tank.'

We tasted samples of Einar's proto-whisky, both peated and unpeated from sherry and bourbon casks. His whisky was three years old in 2019, but like most Faroese he is not in a hurry and will release his whisky to the world when the time is right.[9] Maturation in wood takes time in Klaksvík, which, like the rest of the Faroes, has an extreme oceanic climate with only a small difference between average winter and summer temperatures. As we say our farewells, I tell Einar of that often-repeated Scottish proverb 'Today's rain is tomorrow's whisky'. He smiles. Watching the rain and Faroese patience will be the name of the game.

Back in Tórshavn that evening, the rain has gone and the sun is out. It feels like midsummer again and we take a walk to clear our heads after sampling the spirit of Klaksvík. The dying embers of a huge bonfire on the beach, fully clothed and yellow-wellied children splash happily in the sea, smiling parents look on and everything bathed in that sharp evening light that is so special in the northern countries at this time of year. A young family of eider ducks float gently by on a peat-calm sea, working the rocky shoreline, communicating softly to each other with their mildly surprised, comical calls. Tonight, at 62°N, there is a mood of peace and calm – the midsummer sun sets at 11.22pm.

Acknowledgements

First and foremost, huge thanks to Neil Wilson, who, from seeing the first chapters, was unwavering in his support for the project and acted as an experienced guide and mentor. He not only edited an earlier version of the manuscript but helped to find a route to publication through his contacts, knowledge and understanding of the literary and whisky world. Special thanks to Dave Broom who contributed both a compelling introduction to the book and his own insights as a liquid antiquarian.

Researching *Peat and Whisky* was a journey, sometimes an odyssey, that took me to expected and unexpected places. The book is also about people and owes much to the individuals who made this project both enlightening and fun. All of the following, in no particular order, have contributed in their own way. Apologies if I have failed to recognise your contribution.

Starting at Knockdhu Distillery with Gordon Bruce, the whisky world has been open, welcoming and receptive to my intrusions and questions. Iain McArthur (Lagavulin and former employee of Port Ellen Distillery) told me about the old ways on Islay; Derek Scott (Kilchoman) showed me how it was done. John Thomson, whose willingness to impart his knowledge and insight into distilling and malting was truly educational. In Campbeltown, Iain McAlister (Glen Scotia), Kate and Mark Watt (Watt Whisky) and Springbank people John, Roddy, Findlay Ross and Lea Watson were welcoming and willing guides. Brian Kinsman introduced me to the world of blending and Angus MacRaild talked old-style whisky. On a wet day in Skye Bruce Perry, Neil Mathieson and the team at Torabhaig turned it into a special distillery visit. Prof. Alan Wolstenholme, Dr Nick Morgan and Alan Winchester kept facts straight, while Marcel van Gils and Josh Feldman helped in my exploration of the story of

PEAT AND WHISKY

Laphroaig. Mark Davidson (aka the Jolly Topper) is always a font of knowledge as well as a font of whisky. Thanks to Billy Abbott, Charles MacLean MBE, Gavin D Smith and Brian Townsend for their enthusiasm and support of the book.

Away from the shores of Scotland, Chris Middleton provided valuable insights into the links between Scotch whisky and Australia; Matt Hoffman and Jason Parker (Westland and Copperworks Distilleries) spoke of exciting times ahead in the Pacific Northwest. Einar Waag was a wonderful host on a filthy day in the Faroe Islands; Peter Mulryan and his team at Blackwater Distillery were equally welcoming and equally passionate about their spirit. Fredrik Svärdell (Mackmyra), and Ashok Chokalingam (Amrut) answered my questions about Swedish and Indian whisky.

The book draws on discussions, meetings and the hospitality of many non-whisky people that I met on my travels. To mention just a few: Lorraine and Billy at the St Laurence Bar (Slamannan), ex-miners James Burns Hogg and Neil Young, Campbeltown-based historian Angus Martin, Dr Marcus Wallin in Uppsala, Lesley Craig (Northern Peat & Moss Company), Jimmy FitzPatrick at Carron Lodge. Ken Galloway, George MacDonald, Donna Fairbairn and Sandy Matheson all helped to bring back to life the story of the Old Chivas Regal site on Lewis. The Altnaharra Hotel, Kylesku Hotel, Clynelish Farmhouse and Garavan's Bar in Galway provided a warm and welcoming sanctuary at important moments.

From the academic world, Prof. Roxane Andersen (University of Highlands and Islands) and Prof. Nigel Roulet (McGill University, Montreal) were willing accomplices. Prof. Kevin Edwards (University of Aberdeen) provided valuable insights into Faroe Islands peat. Dr Julian South, Executive Director of the Maltsters' Association of Great Britain (MAGB) opened my eyes to barley and Dr Barry Harrison (SWRI) was a wonderful guide into the world of phenols.

ACKNOWLEDGEMENTS

A special mention to Nicky Melville, teacher of Creative Writing at the University of Edinburgh, who helped me break away from the shackles of scientific writing and later in life encouraged me to express myself and write in a different way.

The book owes much to the archives, archivists, libraries and librarians who open doors to the past. I'd like to thank Andrea Massey at the Orkney Library & Archive, Niki Russell of the Archives & Special Collections at the University of Glasgow Library, Nicola Moss, Curator at the National Mining Museum Scotland, Joanne McKerchar, Diageo Archive Menstrie, and numerous staff at the National Library of Scotland in Edinburgh for digging deep and finding much. Carol Quinn, Head of Archives at Irish Distillers, not only righted a few wrongs, but was a wonderful guide into the fascinating history of Irish whiskey. The British Newspaper Archive is a rich resource and never failed to add substance to my writing.

Thanks to willing friends who read and provided valuable feedback on early versions of chapters: Debbie and Frank Harvey, Alan and Julia Taylor, David and Sheila Fathers, Anne and Stuart Craig and the self-styled Geolly Boys (you know who you are). It is always a Team effort.

Special thanks to my long-suffering brother and handy paramedic Bob, for not only reading my words, but also in a moment of sheer madness, agreeing to spend a week walking across ninety percent water. He was a more willing accomplice on our pre-pandemic road trip to the bogs of Ireland. We have history. But most importantly, I would like to thank Sofia, who apart from myself, was my toughest and most constructive critic. She read all that I wrote, came up with the best chapter title and never once doubted that *Peat and Whisky* would find a way.

A special thanks to Sara Hunt for her passion for the book and for supporting me through the publishing process. It has been a pleasure to work with Sara and her team at Saraband: editor Heather Merrick, Rosie Hilton and Ellie Croston. Thank you all.

And finally, to all those wonderful people who promised to buy a copy of *Peat and Whisky – The Unbreakable Bond*. I hope it was worth the wait. *Slàinte Mhath!*

Author's Note

This manuscript relies on significant amounts of source material and although it has been checked for errors and omissions, no work of this type is perfect. Personal quotes were rechecked with individuals wherever possible. Apologies if I have got something wrong; I take full responsibility.

Notes and References

Chapter 1

1. *Whisky Island*, Scottish Television Film Documentary, 1965
   https://www.youtube.com/watch?v=TbPO7PXEkPE

2. *Huffpost*, Whisky Tourism - The Impact On A Scottish Island,
   14.07.2017

3. David Webster and others, *A Guide to the Geology of Islay*,
   Ringwood, Glasgow, 2015, p.39

4. 15,000 years BP (before present) is equivalent to 13,000 years
   BCE (before the common era) or 13,000 years BC (before
   Christ). The end of the ice age and start of the Holocene period
   is dated at 11,700BP

5. Nicholas Morgan and the Whisky Exchange, *Everything You
   Need to Know About Whisky (but are too afraid to ask)*, Ebury
   Press, 2021, p.231

6. SMWS America website, Cask Curriculum: Peat Origins,
   https://www.smwsa.com/blogs/cask-curriculum/peat-origins

7. EA Martin, *The Story of a Piece of Coal*, George Newnes Ltd,
   London, 1896

8. Arthur Holmes, *Principles of Physical Geology*, Nelson, London,
   1944, p.441, p.437

9. Ali Whiteford, *An Enormous Reckless Blunder*, IBT, 2017, p.1, p.63

10. GK Fraser, *Peat Deposits of Scotland*, Wartime Pamplet No. 36,
    Part 1, 1943

11. *Trends in Atmospheric Carbon Dioxide*, www.esrl.noaa.gov/gmd/
    ccgg/trends

12. William King, *Of the Bogs, and Loughs of Ireland*, Phil. Trans.
    Royal Soc. Lond., 1685

13. At the time "*Torys*" in Ireland were guerrillas loyal to the
    Royalist supporters of Union bandits – derived from the Gaelic,
    toiraidh – pursued, or men on the run

14. Nicholas Turner, *An Essay on Draining and Improving Bogs in
    which the Nature and Properties are Fully Considered*, Baldwin &
    Brew, London, 1784, p.1

15. *Soil Map of the World, 1:5,000,000*, Volume 1, Legend, Food
    and Agriculture Organisation (FAO) of the United Nations,
    UNESCO, Paris, 1974

16. L von Post, *Das Genetische System der Organogenen Bildungen Schwedens*, Comité International de Pédologie, IV Commission. No. 22, 1924

17. Charles Darwin, *The Beagle Diary*, March 1833, darwin-online. org.uk

18. Joseph D Hooker, *Correspondence from the Antarctic Expedition, 5th April 1842*, Royal Botanic Gardens, Kew, https://jdhooker. kew.org/p/jdh

19. Wilson, P and others, *Soil erosion in the Falkland Islands: an assessment*, Applied Geography 13, 1993, p.329-352,

20. https://www.theflowcountry.org.uk

21. GC Dargie and others, *Age, extent and carbon storage of the central Congo Basin peatland complex. Nature*, 542, 2017, p.86-90

22. Dan Charman, *Peatlands and Environmental Change*, John Wiley & Sons, 2002

23. Osgood Mackenzie, *A Hundred Years in the Highlands*, 1921, National Trust of Scotland 1988 edition, p.214, p.124

24. Alfred Barnard, *The Whisky Distilleries of the United Kingdom*, Harper's Weekly Gazette, London, 1887, Birlinn 2012 reprint

25. WA Kerr, *Peat and Its Products*, Begg, Kennedy & Elder, Glasgow, 1905, p.181

26. Gavin Smith, *The Whisky Men*, Birlinn, 2005, p.70

27. James Murray, *Jim Murray's Complete Book of Whisky*, Carlton, 1997, p.84

28. *Aberdeen Weekly Journal*, Visit to Peat and Moss Litter Factory, 06.07.1917

29. Testimony of Dr Jeffray, "Agent and Traveller" for the Haig family who at the time owned Lochrin Distillery in Edinburgh. *Report Respecting the Scotch Distilling Duties*, House of Commons Papers printed 12th July 1799, Appendix M, p.370

Chapter 2

1. James Duff, *Collection of Poems, Songs etc*, Morison, Perth, 1816, p.115

2. K Branigan and others, *Bronze Age fuel: The oldest direct evidence for deep peat cutting and stack construction? Antiquity*, 2002, 76, p.849-855

NOTES AND REFERENCES

3. *Orkneyinga Saga*, Penguin Books, 1981, p.29

4. https://www.undiscoveredscotland.co.uk/burghead/fort/index.html

5. Robin Crawford, *Into the Peatlands*, Birlinn, 2018, p.189, p.194

6. Donald S. Murray, *The Dark Stuff*, Bloomsbury, 2018, p.64, p.54

7. MAW Gerding and others, *The history of the peat manufacturing industry in The Netherlands: Peat moss litter and active carbon*, Mires and Peat, 2015, 16, p.1-19

8. https://www.lowtechmagazine.com/2011/09/peat-and-coal-fossil-fuels-in-pre-industrial-times.html

9. Vincent van Gogh, *Women on the Peat Moor*, Van Gogh Museum, Amsterdam, 1883

10. Robert Moray, *An Account of the Manner of Making Malt*, Phil. Trans. Royal Soc. Lond., 1665-1678, 12, p.1069-1071

11. Martin Martin, *A Description of the Western Isles of Scotland circa 1695*, Birlinn 1994 edition

12. Thomas Pennant, *A Tour in Scotland and Voyage to the Hebrides 1772*, Birlinn 1998 edition

13. Samuel Johnson and James Boswell, *A Journey to the Western Islands of Scotland*, originally published 1785, Oxford University Press 1984 edition

14. Joseph Mitchell, *Reminiscences of My Life in the Highlands*, Vol 1-2, originally published 1883-84, David and Charles, Newton Abbot 1971 edition

15. John MacCulloch, *The Highlands and Western Isles of Scotland*, Vol 1-4, Longman, Hurst, Rees, Orme, Brown, and Green, London, 1824

16. Archibald Geikie, *Scottish Reminiscences*, James Maclehose & Sons, Glasgow, 1904, and p.266

17. Alfred Barnard, *The Whisky Distilleries of the United Kingdom*, Harper's Weekly Gazette, London, 1887, Birlinn reprint 2012

18. Compton MacKenzie, *Whisky Galore*, Chatto and Windus, London, 1947, p.139, p.144, p.123

19. Comment in 1832 attributed to Lewis Cumming from an article written by Richard Woodward, 03/10/2019, in the online magazine Scotchwhisky.com

20. JA Nettleton, *The Manufacture of Whisky and Plain Spirit*, Cornwall and Sons, Aberdeen, 1913

21. Testimony of Mr Frederick Maclagan, "Agent and Traveller" for the Haig family who at the time owned Lochrin Distillery in Edinburgh. *Report Respecting the Scotch Distilling Duties*, House of Commons Papers v119, printed 11.06. 1798, p.306, p.308,

22. Osgood Mackenzie, *A Hundred Years in the Highlands*, first published 1921; (National Trust of Scotland 1988 edition), p.173

23. Steven Sillett, *Illicit Scotch*, Beaver Books, 1965, p.85

24. *Jamieson's - An Etymological Dictionary of the Scottish Language*, 1880

25. *The Amber Light*, a documentary film written by David Broom, directed by Adam Park, 2019

26. Inge Russell and Graham Stewart (eds), *Whisky – Technology, Production and Marketing*, 2nd edition, Elsevier, 2014, p.8

27. James Murray, *Jim Murray's Complete Book of Whisky*, Carlton, 1997, p.110, p.58, p.40, p.56, p.38

28. Dating back to the 14th century, 'Beyond the Pale' became a widely used expression, meaning beyond the bounds of acceptable behaviour

29. Peter Mulryan, *The Whiskeys of Ireland*, The O'Brien Press, Dublin, 2002, p.30

30. Michael Donovan, *Domestic Economy*, Longman, Rees, Orme, Brown, Green and Taylor, London, 1830, p.253

31. Glenbuchat Heritage Archive, http://www.glenbuchatheritage.com

32. *Matuara Ensign*, The Tarland Cenetarian, 15, 14.04.1893

33. Charles MacLean and Daniel MacCannell, *Scotland's Secret History*, Birlinn, 2017, p.36

34. Gavin Smith, *The Secret Still*, Birlinn, 2002, p.87

35. David Buchan and James Moreira (Eds), *The Glenbuchat Ballads*, The University Press of Mississippi, 2007

36. *The Second or "New" Statistical Account of Scotland*, Vol.12, Parish of Glenbucket, authored by Rev. Robert Scott, Blackwoods and Sons, 1834-45, p.437-38

37. Glenlivet Distilling Company, *A Taste of Heaven Before the Trumpets Blow*, trade advert, 1989

NOTES AND REFERENCES

38. Aeneas MacDonald, *Whisky*, Porpoise Press, 1930, p.88, p.87

39. JG Phillips, *Wanderings in the Highlands of Banff and Aberdeen Shires*, Banffshire Journal Office, 1881, p.17

40. *The Sphere*, 23.08.1958

41. *Dundee Advertiser,* The Highest Village in Scotland, 05.10.1895

42. www.whiskyfun.com/archivedecember09-2.html#211209

43. Neil Gunn, *Whisky and Scotland*, George Routledge & Sons, 1935, p.183

44. John Hulme and Michael Moss, *The Making of Scotch Whisky*, Canongate, 1981, p.49

45. https://whiskyauctioneer.com/lot/119846/macallan-1946-select-reserve-52-year-old

46. Rare Whisky 101, *Half Year Review 2018*, https://www.rarewhisky101.com

47. Michael Jackson, *World Guide to Whisky*, 1987, p.61

48. Scottish Rights of Way and Access Society, *Mannoch Road*, www.heritagepaths.co.uk

49. *Aberdeen Press and Journal,* William Russell's Passing Glances, 23.11.1949

50. H. Charles Craig, *The Scotch Whisky Industrial Record*, citing the Minutes of the September 1967 meeting of the Longmorn-Glenlivet Distillers Ltd Board, 1994, p.300

51. Robin Laing, *The Whisky River*, Luath Press, Edinburgh, 2007, p.57-58

52. The Glenrothes Distillery website, https://www.theglenrothes.com/en/history

53. The Balmenach Distillery, www.discover-secret-scotland.co.uk/whisky-reviews/balmenach.php

54. PJA Burt, *The Great Storm and the fall of the first Tay Rail Bridge*, J. Royal Met. Soc., doi: 10.1256/wea.199.04, 2004

55. *Eday Peat Company records1926-1965*, D1/4, Orkney Archive, Kirkwall

56. Eday – The Isthmus Isle, https://www.orkneyguide.com/ogbpdf/Eday.pdf

57. Rosemary Hebden, *Eday – Orkney's Best-Kept Secret*, Carrick Press, Eday, 2008,

58. *London Courier and Evening Gazette*, Island of Eday, Orkney in north Scotland for Sale by Private Bargin, 25.07.1834

59. *Orkney Herald*, Peats for Export, 17.04.1929

60. *The Scotsman*, Well-known Orkney steamer laid up, 02.12.1933

61. *Banff Journal and General Adviser*, Feat of strength and ingenuity, 31.03.1863

Chapter 3

1. Alfred Barnard, *The Whisky Distilleries of the United Kingdom*, *Harper's Weekly Gazette*, London, 1887 (Birlinn reprint 2012), p.330

2. Gavin D Smith, *The Whisky Men*, Birlinn, 2005, p.252

3. Brian Townsend, *Scotch Missed*, Neil Wilson Publishing Ltd, 2015, p.130

4. *Falkirk Herald and the Linlithgow Journal*, Sketch of the St. Magdalene Distillery, sourced from the *Scottish Standard*, 30.06.1870

5. *Scotsman*, St. Magdalene's Malt Distillery, Linlithgow, For Sale, 5.09.1874

6. https://www.hutton.ac.uk/learning/natural-resource-datasets/peat-surveys/peat-deposits

7. *Falkirk Herald and Linlithgow Journal*, Fannyside Muir - Peat versus Coal, 05.08.1872

8. https://malt-review.com/2019/03/18/rosebank-vertical/

9. Why some distilleries use fire heated stills. sw.com, 10.09.2019

10. *Dundee Courier*, Peat to Drive the Herring Boats, 10.09.1937

11. WA Kerr, *Peat and Its Products*, Begg, Kennedy & Elder, Glasgow, 1905, p.21

12. Brian Ashcraft, *Japanese Whisky*, Tuttle Publishing, 2018, p.139

13. The Carron Company: An Introduction, www.ourstoriesfalkirk.com/

14. ED Hyde, *Coal Mining in Scotland*, Scottish Mining Museum, 1987

15. MK Oglethorpe, *Scottish Collieries*, Historic Environment Scotland, 2006, p.43.

NOTES AND REFERENCES

16. Con Gillen, *Geology and Landscapes of Scotland*, 2nd edition, Dunedin, 2013, p.202

17. Archibald Geikie, *Scottish Reminiscences*, James Maclehose & Sons, Glasgow, 1904, p.102

18. Joseph Mitchell, *Reminiscences of My Life in the Highlands*, Vol 2, originally published 1883-84; David and Charles, Newton Abbot 1971 edition, p.162

19. JA Nettleton, *The Manufacture of Whisky and Plain Spirit*, Cornwall and Sons, Aberdeen, 1913

20. Ali Whiteford, *An Enormous Reckless Blunder*, IBT, 2017, p.72

21. Angus Martin, *Kintyre Country Life*, John Donald Publishers, Edinburgh, 1987, p.93

22. Federal Distillery, Oz Whisky Review, www.ozwhiskyreview. com.au/federal

23. www.islayinfo.com/dougie-macdougall-port-askaig-islay

24. Bunnahabhain coal book, GB248 UGD 18/8 - Series, Coal Books, 1928-64, Archives and Special Collections - University of Glasgow Library

25. Jim Murray, *Whisky Bible 2017*, Dram Good Books Ltd, 2016, p.94

26. www.malts.com/en-gb/products/single-malt-whisky/brora-30-year-old-2006

27. David Stirk, *The Distilleries of Campbeltown: The Rise and Fall of the Whisky Capital of the World*, Angel's Share, 2007, p.61

28. Angus Martin, *Kintyre Country Life*, John Donald Publishers, Edinburgh, p.90, p.92

29. Nicholas Morgan, *A Long Stride*, Canongate, 2020, p.39

30. http://www.springbank.scot/about/story/

31. RA Herd and AG Wolstenholme, *On the Production Methods of Pot Still Whisky, Campbeltown Scotland, May 1920 Masataka Taketsuru*, Humming Earth, Edinburgh, 2021, p.17

32. Carl Lönndahl, *Between yeast and stills: the story of Sweden's first whisky, A documentary*, Vipper Media AB, 2018, https://www.youtube.com/watch?v=SNt8LeAF_Wc

33. Thomas Pennant, *A Tour in Scotland and Voyage to the Hebrides 1772*, Birlinn, 1998 edition, p.184

34. John MacCulloch, *The Highlands and Western Isles of Scotland*, Vol 4, Longman, Hurst, Rees, Orme, Brown, and Green, London, 1824, p.371

35. David C Kerr, *Shale Oil Scotland: the World's Pioneering Oil Industry*, The Author, 1994, p.56

36. H Charles Craig, *The Scotch Whisky Industrial Record*, Index Publishing Ltd, 1994, p.368

37. Peter Spiller, *Cardhu - The World of Malt Whisky*, JW & Sons, London, 1985

38. Difford's Guide - Craigellachie Distillery, www.diffordsguide.com/producers/

39. JS Owen, *Coal Mining at Brora 1529-1974*, The Highland Council, 1995, p.1, p.3

40. Patrick Brossard, *Brora - A Legendary Distillery (1819-1983) and Whisky*, 2006, p.61

41. Marx, Karl. *Capital. Volume 1.* (1867), New York: International Publishers, New York, 1966

Chapter 4

1. P Schidrowitz, *The Application of Science to the Manufacture of Whisky*, J. Inst. Brewing, 1907, p.159-176

2. E Ledesma E and others, 'Smoked Food, Current Developments in Biotechnology and Bioengineering', In: *Food and Beverages Industry*, Elsevier, 2017, p.201-243

3. *Norfolk Chronicle*, shop advertisement, 03.04.1880

4. *The Rotarian, Unusual Stories of Unusual Men:* Ernest H. Wright — Classification: "Condensed Smoke", 1923, 240, p.209–210

5. EA Gomaa and others, *Polycyclic aromatic hydrocarbons in smoked food products and commercial liquid smoke flavourings*, Food Additives and Contam., 1993, 10, p. 503-521

6. https://www.kingdomscotland.com

7. Samuel Morewood, *An essay on the inventions and customs of both ancients and moderns in the use of inebriating liquors. Interspersed with interesting anecdotes, illustrative of the manners and habits of the principal nations of the world*, Longman, London, 1824, p.64

NOTES AND REFERENCES

8. Kate Winkler, *Death in the Air: The True Story of a Serial Killer, the Great London Smog, and the Strangling of a City*, Hachette Book Group, 2017

9. Schlenkerla – the historic smokebeer brewery, https://www.schlenkerla.de/indexe.html

10. C Da Porto and others, *A study on the composition of distillates obtained from smoked marc*, Anal. Chim. Acta, 2006, 563, p.396-400

11. *Wine Magazine*, Breaking Down the Difference Between Mezcal and Tequila, https://www.winemag.com/2019/08/27/difference-mezcal-vs-tequila/

12. Neil M Gunn, *Whisky and Scotland*, George Routledge and Sons, 1935, p.133

13. Michael Jackson, *World Guide to Whisky*, 1987, p.36, p.184

14. SH Hastie, *Character in Pot Still Whisky*, J. Inst. Brewing, 1926, p.209-220

15. *New York Times*, Jack Daniel's Embraces a Hidden Ingredient: Help From a Slave, 2016, https://www.nytimes.com/2016/06/26/dining/jack-daniels-whiskey-nearis-green-slave.html

16. P Schhidrowitz and F Kaye, *The Distillation of Whisky*, J. Inst. Brewing, 1906, p.496-517

17. www.westcorkdistillers.com

18. Ian Buxton and Paul Hughes, *The Science and Commence of Whisky*, RSC Publishing, 2014, p.251

19. *New Scientist*, Largest ever wildfire in Greenland seen burning from space, 08.08.2017

20. LL Bourgeau-Chavez, *Assessing Boreal Peat Fire Severity and Vulnerability of Peatlands to Early Season Wildland Fire*, Front. in For. and Glob. Change, https://doi.org/10.3389/ffgc.2020.00020

21. G Rein, *Smouldering Fires and Natural Fuels*, In: Fire Phenomena in the Earth System – An Interdisciplinary Approach to Fire Science, Belcher (ed), Wiley and Sons, 2013, p.15–34

22. *Dalmore Distillery's Fire Heritage*, Fire Cover, 229, 08.2020

23. Photo of the gutted remains of the Dalmore Distillery peat shed 1919, canmore.org.uk/collection/1445225

24. *Upsala Nya Tidning, Här går 100 miljoner upp i rök,* 17.11.1990

25. BBC News, *Huge Flow Country wildfire doubled Scotland's emissions,* 18.11.2019

26. CS McBain, *A Pioneer of the Spirit – Charles Cree Doig,* Scottish Industrial History, **20**, 2000, Business Archives Council of Scotland, p.47-51

27. John Thomson, *An Investigation of Factors Influencing Efficient Phenol Production in Highly Peated Distilling Malt,* unpublished thesis for MAGB Malting Diploma, 2004

28. Iain Russell, *A History of Glenlivet (the place),* Edinburgh Whisky Academy, Online resource

29. James Murray, *Jim Murray's Complete Book of Whisky,* Carlton, 1997, p.85

30. Gavin Smith and Graeme Wallace, *Ardbeg – A Peaty Provenance,* GW Publishing, 2008, p.139, p.109

31. New Whiskies, Batch 196, tasting notes, Dave Broom, sw.com, 12.04.2019

32. *guga* is young dried gannet – a delicacy for some, a horror to others – that is produced each year after a boat full of Lewismen return from their annual harvesting trip to the uninhabited island of Sula Sgeir in the North Atlantic

33. JA Nettleton, *The Manufacture of Whisky and Plain Spirit,* Cornwall and Sons, Aberdeen, 1913, p.583

34. Robert Moray, *An Account of the Manner of Making Malt,* Phil. Trans. Royal Soc. Lond., 1665-1678, 12, p.1069-1071

35. Derek Cooper, *The Century Companion to Whiskies,* Century, 1983, p.114, p.127

36. Misako Udo, *The Scottish Whisky Distilleries,* Distillery Cat Publishing, 2005, p.226

37. John Thomson, *Port Ellen Distillery and Maltings,* first published 2007, Diageo, 2018

38. GN Bathgate, *The influence of malt and wort processing on spirit character: the lost styles of Scotch malt whisky,* J. Inst. Brewing, 2019, 125, p.200-213

39. AT Lucas, *Toghers or Causeways: Some Evidence from Archæological, Literary, Historical and Place-Name Sources,* Proc. of the Royal Irish Academy: Arch., Culture, Hist., Lit., 1985, 85, p.37-60

NOTES AND REFERENCES

40. Joseph Mitchell, *Reminiscences of My Life in the Highlands*, Vol 1, originally published 1883-84; David and Charles, Newton Abbot 1971 edition, p.198

41. Historic Irish Moonshine Found in Tyrone, Whisky Online Blog, 12.12.2019

42. WA Kerr, *Peat and its Products: An Illustrated Treatise on Peat and Its Products as a National Source of Wealth*, Beg, Kennedy and Elder, Glasgow, 1905, p.229

43. *Smithsonian Magazine*, How Humble Moss Healed the Wounds of Thousands in World War I, 28.04.2017

44. Robin Wall Kimmerer, *Gathering Moss: A Natural and Cultural History of Mosses*, Oregon State University Press, 2003, p.105-7, p.113

45. PG Ayres, *Isaac Bayley Balfour, Sphagnum moss, and the Great War (1914–1918)*, Arch. of Nat. Hist., 2015, 42, p.1-9

46. Chris Middleton, personal communication

47. Gillian Brown, *Mosses, liverworts and hornworts Significant bryophyte collections at the University of Melbourne Herbarium*, University of Melbourne Collections, Issue 8, 06.2011

48. JCS Kleinjans and others, *Polycyclic aromatic hydrocarbons in whiskies*, The Lancet, 348, 21.12.1996

49. William King, *Of the Bogs, and Loughs of Ireland*, Phil. Trans. Royal Soc. Lond., 1685

50. Gavin Smith, *The Secret Still: Scotland's Clandestine Whisky Makers*, Birlinn, 2002, p.159

51. *Lewisiana, or Life in the Outer Hebrides*. William Anderson Smith, London, 1875

52. Ali Whiteford, *An Enormous Reckless Blunder*, IBT, 2017, p.63, p.62

53. Dr Weber's Croonian Lectures, Correspondence to the Editor, *The Lancet*, 05.1885, p.821

54. *Medical and Surgical Reporter*, The March of the Epidemic, 04.05.1872, 26, p.18

55. Alexander Fleming, News Summaries, 22.03.1954

56. BBC Tonight Programme, *Flu Cures*, Reporters Alan Whicker and Fyfe Roberston, 18.02.1959

57. *Scottish Field*, Prime Ministers and their Relationship with a Dram, 10.03.2018

58. *The Amber Light*, a documentary film written by David Broom, directed by Adam Park, 2019

59. Raphael Holinsted, *History of Ireland*, 1577

60. Martin Martin, *A Description of the Western Isles of Scotland circa 1695*, Birlinn 1994 edition, p.144, p.220

61. Archibald Geikie, *Scottish Reminiscences*, James Maclehose & Sons, Glasgow, 1904, p.404

62. *Belfast Newsletter,* Bushmills advert, 15.03.1940

63. *Belfast Telegraph,* Jameson whiskey advert, 03.01.1957

64. D Clark, *The rise and demise of the "Brompton Cocktail"*, In: ML Meldrum (ed) *Opioids and Pain Relief: a Historical perspective.* Progress in Pain Research and Management, 2003, 25, IASP Press, Seattle, p.85-98

65. *British Medical Journal,* Whisky, 26.12.1903, p.1645-1651

66. RB Weir, *The History of the Malt Distillers Association of Scotland*, Elgin, 1974, p.71

67. Imperial War Museum, *Lives of the First World War, Stuart Henderson Hastie*, https://livesofthefirstworldwar.iwm.org.uk/lifestory/1612632

68. Gregory H Millar, *Whisky Science: A Condensed History*, Springer, 2019, p.173

69. RA Herd and AG Wolstenholme, *On the Production Methods of Pot Still Whisky, Campbeltown Scotland, May 1920 Masataka Taketsuru*, Humming Earth, Edinburgh, 2021

70. RB Weir, *The History of the Distiller's Company*, 1995, Clarendon Press, p.86

71. SH Hastie, *From Burn to Bottle*, Scotch Whisky Association, 3rd impression, 1956

72. Distiller's Company Limited, *Gazette*, 10.1926

73. C Macfarlane, *The Estimation and identification of phenols in malt from peat-fired kilns and some applications of the analysis*, J. Inst. Brewing, 74, 1968, p.272-275

74. The National Archives, Muntons PLC, Stowmarket, HC466, https://www.nationalarchives.gov.uk

75. Marcel van Gils and Hans Offringa, *The Legend of Laphroaig*, Exhibitions International, 2008, p.81

NOTES AND REFERENCES

76. D Howie D and JS Swan, *Compounds influencing peatiness in Scotch malt Whisky flavour.* In: Nykänen L, Lehtonen P (eds) Proceedings of the Alko symposium on flavour research of alcoholic beverages. Helsinki, Foundation for Biotechnical and Industrial Fermentation Research, Helsinki, 1984, p.279–290

77. DD Singer and JW Stiles, *The Determination of Higher Alcohols in Potable Spirits: Comparison of Colorimetric and Gaschromatographic Methods*, The Analyist, 1965, 90, p.290-296

78. Richard Grindal, *The Spirit of Scotland*, Warner Books, 1992, p.216

79. C Macfarlane and others, *The Qualitative Composition of Peat Smoke*, J. Inst. Brewing, 79, 1973, p.202-209

80. ME Stevens and others, *On the Expectorant Action of Creosote and the Guaiacols, Can. Med. Assoc. J., 1943, 48, p.124–127*

81. *A Influenza Hespanola,* Advisory to the People of Brazil from the Inspector of Hygiene, https://www.diariodorio.com/historia, 18.03.2020

82. Chemistry World Podcast, *Guaiacol*, 04.12.2013 www.chemistryworld.com/podcats/guaiacol/6861.article

83. BM Harrison and others, *Differentiation of Peats Used in the Preparation of Malt for Scotch Whisky Production Using Fourier Transform Infrared Spectroscopy*, J. Inst. Brewing, 2006, 112, p.333–339

84. BM Harrison and FG Priest, *Composition of Peats Used in the Preparation of Malt for Scotch Whisky Productions – Influence of Geographical Source and Extraction Depth*, J. Agric. Food Chem., 2009, 57, p.2385–91

85. Neville Peat, *Shackleton's Whisky*, Preface, 2013, p.81

86. J Pryde and others, *Sensory and Chemical Analysis of 'Shackleton's' Mackinlay Scotch Whisky*, J. Inst Brewing, 2011, 117, p.156-65

87. Tasting notes on Mackinlay's Shackleton whisky, www.whiskybase.com

88. *Illustrated London News*, Laphroaig Distinctly Different Whisky, 01.11.1997

89. Melnick Medical Museum Blog, *Medicinal Alcohol and Prohibition*, 07.04.2010

90. Leung A and others, *History of U.S. Iodine Fortification and Supplementation*, Nutrients, 2012, 4, p.1740-46

91. JH Lazarus and others, *Iodine in Malt Whisky: A Preliminary Analysis*, Thyroid, 2017, 2017, 27, p.477-478

92. P Bendig and others, *Quantification of Bromophenols in Islay Whiskies*, J. Agric. Food Chem., 2014, 62, p.2767–71

93. *Boston Daily Advertiser*, Ilay whisky story, 1926, cited in SH Hastie, *Character in Pot Still Whisky*, J. Inst. Brewing, 1926, p.210

94. JR Hulme and MS Moss, *The Making of Scotch Whisky*, Canongate, 1981, p.173

Chapter 5

1. Philip Morrice, *"Schweppes Guide to Scotch"*, Alphabooks, 1983, p.19

2. *Aberdeen Press & Journal,* Distilleries Face Another Drop in Production, 19.02.1968

3. The Maltsters Association of Great Britain, https://www.ukmalt.com/wp-content/uploads/2021/01/maltings-sites-2021-PDF.pdf

4. The Middle Cut: A Whisky Podcast, *Let's Talk Caol Ila*, Ali Reynolds and Colin Dunn, Apple Podcats

5. Andrew Jefford, *Peat, Smoke and Spirit; a Portrait of Islay and its Whiskies*, Headline, 2004, p.178

6. Gavin D Smith, *The Whisky Men*, Birlinn, 2005, p.199

7. Whisky Exchange Blog, *How do you make the World's Peatiest Whisky?*, 15.12.2017

8. M Kyraleou and others, *The Impact of Terroir on the Flavour of Single Malt Whisk(e)y New Make Spirit*, Foods, 2021, 10, 443, https://doi.org/10.3390/foods10020443

9. Virtual Whisky Show 2020, *Whisky Terroir: A Con?*

10. Rob Arnold, *The Terroir of Whiskey*, Columbia University Press, 2021, p.6

11. Archibald Geikie, *Scottish Reminiscences*, James Maclehose & Sons, Glasgow, 1904, p.114

12. SH Hastie, *Character in Pot Still Whisky*, J. Inst. Brewing, 1926, p.209-220

13. Aeneas MacDonald, *Whisky*, Porpoise Press, 1930, p.79

14. Margaret Fay Shaw, *Folksongs and Folklore of South Uist*, Birlinn, 2014

15. B Harrison, *Peat Source and its Impact on the Flavour of Scotch Whisky*, PhD Thesis, Heriot-Watt University, 2007

16. Bunnahabhain Peat Cuttings Books, GB248 UGD 218/17, 1885-1960s. Archives and Special Collections, University of Glasgow Library

17. Alfred Barnard, The Whisky Distilleries of the United Kingdom, *Harper's Weekly Gazette*, London, 1887, Birlinn reprint 2012, p.163

18. The Story of Glenmorangie, The Ardbeg Project, www.ardbegproject.com/glenmorangie/gmstory.shtml

19. *Aberdeen People's Journal,* New Industry for Aberdeenshire, 03.08.1907

20. *Peterhead Sentinel & General Advertiser for Buchan District* – New Industry for Aberdeenshire – Moss Litter in Pitsligo, 21.09.1907

21. *Aberdeen Press & Journal,* A Novel Export to Australia, 11.06.1909

22. *Buchan Observer & East Aberdeenshire Advertiser,* New Pitsligo Peat for Australia, 27.02.1912

23. *Aberdeen Press & Journal,* Order for Australia, 24.06.1914

24. *Aberdeen Weekly Journal,* Visit to Peat and Moss Litter Factory, 06.07.1917

25. *Aberdeen Press & Journal,* Firm boosted by no-nuke policy, 02.12.1986

26. R Glenworth and JW Muir, *The Soils Round Aberdeen, Inverurie and Fraserburgh*, Soil Survey Memoir, HMSO, 1963

27. Neil Godsman sadly passed away in 2022, three years after my visit to St Fergus Moss, which is described here. The company is now owned by his nephew

28. Svensk Torv, *Torven ger Svensk Whisky Röksmaken*, No.3, 2013

29. http://svenska-industribanor.blogspot.com/2010/01/

30. RF Hammond, *The Peatlands of Ireland*, Soil Survey Bulletin No. 35, An Foras Taluntais, 1981, p.2

31. GB Rigg, *Peat Resources of Washington*, Bulletin No.44, Division of Mines and Geology, State of Washington, 1958

32. *Spokane Review,* New frontier for craft U.S. whiskey may be underfoot in Olympic Peninsula bog, 07.01.2018

33. *Food and Wine,* The Bog-to-Bottle Whiskey Experiment, 24.05.2017

34. Brian Ashcraft and others, *Japanese Whisky,* Tuttle Publishing, 2018

35. Master of Malt Blog, *Tasmanian Whisky – Everything You Need to Know! (Part 4: Belgrove)*, 14.11.2014

36. *Guardian*, For Peat's Sake, 12.11.1977

Chapter 6

1. Dr John Rae, Narrative of an expedition to the shores of the Arctic sea in 1846 and 1847, written on 10th August 1846 on a journey to Repulse Bay, https://archive.org/details/cihm_39502/page/n75/mode/2up

2. *Inverness Courier*, Manufacture of Peat Carbon, 22.10.1874

3. *John O'Groats Journal*, Notes on the Snow Block, 30.03.1876

4. *John O'Groats Journal*, Strathhalladale - Peat for Distilleries, 11.06.1897

5. The Story of Glenmorangie, The Ardbeg Project, www.ardbegproject.com/glenmorangie/gmstory.shtml

6. *Northern Ensign* & *Weekly Gazette,* Through the West Highlands, 02.11.1882

7. Quote sourced from Peter Wright, *Ribbon of Wildness*, Luath Press, Edinburgh, 2010, p.231

8. Frank Fraser Darling, *West Highland survey. An Essay in Human Ecology*. Oxford University Press, 1955

9. *The Scotsman*, Remains of "ancient wolf" found in Highland peat bog, 20.06.2018

10. Nate Silver, *Signal and the Noise: The Art and Science of Prediction*, Penguin, 2013

11. Brian Townsend, *Scotch Missed*, Angels' Share, 2015, p.35

12. Alfred Barnard, *The Whisky Distilleries of the United Kingdom*, Harper's Weekly Gazette, London, 1887 (Birlinn reprint 2012), p.159

13. D Hope and others, *Exports of Organic Carbon in British Rivers*, Hydrological Processes, 1997, 11, p.325-344

14. Caithness Archive Centre, Home Guard/Civil Defence Papers, Report of Bomb Incident at Loch More, 05.06.1940

15. *John O'Groats Journal*, Bank Row, 1940: "Those bombs just came out of the blue", 30.08.2020

16. CK Ballantyne and AM Hall, *The altitude of the last ice sheet in Caithness and east Sutherland, northern Scotland*, Scottish Journal of Geography, 2008, 44, p.169-181

17. RA Lindsay and others, *The Flow Country: the Peatlands of Caithness and Sutherland*, NCC, Peterborough, 1988

18. Andrew Dugmore, *Icelandic volcanic ash in Scotland*, Scottish Geographical Magazine, 1989, 105, p.168-172

19. *Northern Ensign* & *Weekly Gazette*, Altnabreac School Entertainment, 28.08.1879

20. *Aberdeen Press* & *Journal*, School milk - from own cow? 06.02.1969

21. *The Glasgow Herald*, Royal Mail ordered to resume deliveries to remote Highland homes, 31.05.2018

22. *Inverness Courier*, The Story of the Snowed-Up Train at Altnabreac, 15.01.1892

23. *Aberdeen Evening Express*, Power from Peat and Altnabreac to make History, 15.07.1953

24. Scottish Peat Committee, Altnabreac survey, 1950s-60s

25. *Aberdeen Press* & *Journal*, Alarm Over N-Waste Plans, 12.04.1978

26. *The Glasgow Herald*, For Sale at over £700,000…Wilderness with Possibilities, 20.05.1981

27. *The Glasgow Herald*, Caithness Losing its Blanket, 28.08.1995

28. Archibald Geikie, *Scottish Reminiscences*, James Maclehose & Sons, Glasgow, 1904, p133, p.226, p.288

29. DJ Charman, *Blanket mire formation at the Cross Lochs, Sutherland, northern Scotland*, Boreas, 2008, 21, p.53-72

30. Annie Proulx, *Fen, Bog & Swamp*, 4th Estate, 2022, p.84

31. PE Levy and A Gray, *Greenhouse gas balance of a semi-natural peatbog in northern Scotland*, Environmental Research Letters,

2015, 10, 094019

32. DA Ratcliffe, *Mires and bogs*. In: Burnett JH, editor. Vegetation of Scotland. Edinburgh: Oliver and Boyd, 1964, p.426–78

33. *Caithness Courier*, Our First Trip on the Railway to Inverness, 16.07.1880

34. Horace Fairhurst, *Rosal: a Deserted Township in Strath Naver, Sutherland, Proceedings of the Society of Antiquaries Scotland*, 1967-68, 100, p.135-169

35. John MacCulloch, *The Highlands and Western Isles of Scotland*, Vol 1-4, Longman, Hurst, Rees, Orme, Brown, and Green, London, Vol 2, 1824, p.455

36. Dr Isobel F Grant, *Every-day Life on an Old Highland Farm 1769-1782*, Longmans, 1924

37. SW Sillett, *Illicit Scotch*, Beaver Books, 1965, p.101

38. Joseph Mitchell, *Reminiscences of My Life in the Highlands*, Vol 2, originally published 1883-84; David and Charles, Newton Abbot 1971 edition, p.147

39. A transport petition addressed to parliament was then a common way of soliciting for the construction of a new highway

40. Tom Weir, *Land Of Hope And… Golly*, Scots Magazine, https://www.scotsmagazine.com/articles/tom-weir-land-of-hope-and-golly/

41. *The Telegraph*, Duke of Westminster fights to save unique black and white phone box, 08.10.2008

Chapter 7

1. PM Dryburgh, *Scotland's Peat Resources: An Introduction to their Potential*, University of Edinburgh, 1978, p.8

2. Alfred Barnard, *The Whisky Distilleries of the United Kingdom*, Harper's Weekly Gazette, London, 1887, Birlinn reprint 2012, p.160-161, p.544, p.44

3. Patrick Brossard, *Brora: A Legendary Distillery, 1819-1983, and Whisky*, 2016, p.59, p.63

4. Ian Buxton and Paul S Hughes, *The Science and Commerce of Whisky*, Royal Society of Chemistry, 2013, p.108, p.183

5. RM Callender and PF Reeson, *The Scottish Gold Rush 1869*,

NOTES AND REFERENCES

British Mining N.84, 2008

6. Neil Gunn, *Whisky and Scotland*, George Routledge & Sons, 1935, p.157

7. Glenlivet Distilling Company, *A Taste of Heaven Before the Trumpets Blow*, trade advert, 1989

8. Letter to the Dundee Courier, *Export of Scottish Water*, 27.01.1947

9. The Spirit Business, *Polar explorer bottles Port Charlotte with Arctic iceberg water*, 24 July 2017, https://www.thespiritsbusiness.com/2017/07/polar-explorer-bottles-port-charlotte-with-arctic-iceberg-water/

10. Bivrost Whisky, https://www.bivrost.com/whisky/

11. Isfjord, https://www.isfjord.com/

12. BCG Karlsson and R Friedman, *Dilution of whisky – the molecular perspective*, Scientific Reports, 7, 2017, p.6489

13. Malvern Museum, https://malvernmuseum.co.uk/water-cure-room/https://larkfire.com/

14. Whisky Exchange Blog, *Amrut – Nectar of the Gods with Ashok Chockalingam*, 18.10.2018

15. Scotland's People, *Scotland's History: The Ben Nevis Observatory*, https://www.scotlandspeople.gov.uk/article/our-records-ben-nevis-observatory

16. Aeneas MacDonald, *Whisky*, Porpoise Press, 1930, p.83

17. Brian Spiller, *Glenlossie and Mannockmore Distilleries*, DCL Historic Series, 1981

18. Friends of Ben Rinnes, *The lonely grave on Ben Rinnes*, http://www.friendsofbenrinnes.org.uk/Babbies%20Moss.html

19. Alfred Barnard, *The Whisky Distilleries of the United Kingdom*, Harper's Weekly Gazette, London, 1887, Birlinn reprint 2012, p.214

20. Archibald Geikie, *Scottish Reminiscences*, James Maclehose & Sons, Glasgow, 1904, p.153

21. David Daiches, *Scotch Whisky: Its Past and Present*, First published 1969, Birlinn 1995, p.26

22. Scotch Whisky Association, Water Use by the Whisky Industry, 2014

23. Marcel van Gils and Hans Offringa, *The Legend of Laphroaig*, Exhibitions International, 2008, p.81

24. JA Nettleton, *The Manufacture of Whisky and Plain Spirit*, Cornwall and Sons, Aberdeen, p.309

25. SH Hastie, *Character in Pot Still Whisky*, J. Inst. Brewing, 1926, p.209-220

26. Michael Jackson, *World Guide to Whisky*, DK, 1987, p.61

27. Whiskybase, https://www.whiskybase.com/whiskies/whisky/66915/glenisla-1977-sv

28. Helen Bennett, *A murder victim discovered: clothing and other finds from an early 18th-century grave on Arnish Moor, Lewis*, Proceedings of the Society of Antiquaries Scotland, 1977, 106, p.172-182

29. Joseph Mitchell, *Reminiscences of My Life in the Highlands*, Vol 1, originally published 1883-84; David and Charles, Newton Abbot, 1971 edition, p.241

30. Ali Whiteford, *An Enormous Reckless Blunder*, IBT, 2017, p.1, p.63

31. https://www.thewhiskyexchange.com/b/40/craigduff-single-malt-scotch-whisky

32. CS McBain, *Strathisla - 200 years of Distilling Tradition*, Strathisla Distillery, 1986

33. *Daily Mirror*, Drought – Whisky and Water, 14.08.1968

34. *The London Observer*, Drought turns the whisky stills dry, 22 June 2008

35. H Charles Craig, *The Scotch Whisky Industrial Record*, Index Publishing Ltd, 1994, p.317, p.87

36. Nick Bridgland, *Dallus Dhu Distillery*, Historic Scotland, 2002

37. BBC Scotland News, *The twin threats of flooding and drought*, 05.11.2021, https://www.bbc.co.uk/news/uk-scotland-59163739

38. Sir Thomas Dick Lauder, *The Great Floods of August 1829 in the Province of Moray and Adjoining Districts*, published by R Stewart Elgin, 1830

39. Charles MacLean, *World Whisky: A Nation-by-Nation Guide to the Best*, DK Publishing, 2009, p.153

40. Brian Spiller, *Benrinnes Distillery*, DCL Historic Series, 1982

NOTES AND REFERENCES

41. Karl Raitz, Making Bourbon: *A Geographical History of Distilling in Nineteenth-Century Kentucky*, 2020

42. *The Scotsman,* Three rescued from flooded Moray distillery, 08.10.2014

43. J Walker, *Account of the irruption of Solway Moss, on Dec 16, 1772*, Philosophical Transactions, 1772, 62, p.123

44. W.H. Pearsall, *Mountains and Moorlands*, 1950, Collins, London, p.281

45. James Murray, *Jim Murray's Complete Book of Whisky*, Carlton, 1997, p.9

Chapter 8

1. Johnie Walker Trade Advert, *Illustrated London News*, 21.11.1964

2. Patrick Given, *Calico to Whiskey: A Case Study on the Development of the Distilling Industry in the Naas Revenue Collection District 1700-1921*, PhD thesis, National University of Ireland Maynooth, 2011, p.313

3. Malachy Magee, *1000 Years of Irish Whiskey*, The O'Brien Press, Dublin, p.141

4. Advert for Caledonian Peat Reek Whiskey, *Dublin Morning Register*, 10.01.1831

5. Messrs. John Jameson & Sons; William Jameson & Co.; John Power & Son; George Roe & Co, *Truths about Whisky*, Printed by Sutton, Sharpe & Co, London, 1878

6. George Saintsbury, *Notes on a Cellar-Book*, Macmillan, London, 1920, p.79

7. Nicholas Morgan, *A Long Stride*, Canongate, 2020, p.103

8. Alfred Barnard, The Art of Blending Scotch Whisky, *The Whisky Distilleries of the United Kingdom*, Harper's Weekly Gazette, London, 1887, Birlinn reprint 2012, p.542

9. *Daily Mirror,* The flavour of Johnie Walker is 3,000 years old, 9.10.1963

10. Stuart Henderson Hastie, *From Burn to Bottle*, Scotch Whisky Association, 1951, p.14

11. Derek Cooper, *Century Companion to Whiskies*, Century, 1983, p.153

12. Scotch Whisky Industry Review 2019, 42nd edition, Pagoda Scotland Limited

13. Scotch Whisky Industry Review 1983, 6th edition, Campbell Neill & Company

14. *The Guardian,* Geese Feel the Nip of Whisky, 16.07.1985

15. *The Guardian,* Feathers Rustled on Peat Isle, 30.05.1986

16. Nature Conservation Importance of Duich Moss, Scotland, UK, and Potential Damage to Site as a Consequence of Proposed Drainage and Peat Extraction with a Consideration of Possible Alternative Sites, NCC Report to the Secretary of State for Scotland in Nine Parts, 1986, Part 9 Annex 6, Part 7, p.41

17. Andrew Jefford, *Peat, Smoke and Spirit; a Portrait of Islay and its Whiskies*, Headline, 2004, p.353

18. The acrotelm is defined as the upper living and aerobic layer, and the catotelm as the deeper, waterlogged, anaerobic layer, of the peat profile

19. WA Kerr, *Peat and its Products: An Illustrated Treatise on Peat and Its Products as a National Source of Wealth*, Beg, Kennedy and Elder, Glasgow, 1905

20. William King, Of the Bogs, and Loughs of Ireland, *Phil. Trans. Royal Soc. Lond.*, 1685

21. *The Guardian,* End of an era as Ireland closes its peat bogs "to fight climate change", 27.11.2018

22. https://keelingcurve.ucsd.edu/Peat bogs of Québec

23. https://climate.nasa.gov/vital-signs/carbon-dioxide/

24. Sabine Undorf and others, *Learning from the 2018 heatwave in the context of climate change: Are high-temperature extremes important for adaptation in Scotland?*, Environmental Research Letters, 15, 03405, 2020

25. Data for peat extraction, or usage, are either expressed as a weight (tonnes of wet peat, tonnes of dry peat), or as a volume (cubic metres). The last two estimates are based on the most recent (2014) UK mineral extraction data.

26. The Commitment to Responsible Peat Use, Scotch Whisky Association, Summer 2023, www.scotch-whisky.org.uk

27. Office for National Statistics, *UK natural capital: peatlands*, https://www.ons.gov.uk/economy/environmentalaccounts/

NOTES AND REFERENCES

bulletins/uknaturalcapitalforpeatlands/naturalcapitalaccounts

28. Whisky Exchange Blog, *No More Peaty Whisky From Islay*, 01.04.2015

29. *The Glasgow Herald*, Concerns over future of whisky's special ingredient, 03.02.2020

30. *The Independent*, Consultation launched on banning sale of peat in Scotland, 18.02.2023

31. *Sunday Post*, Caithness Sends Peat to Newcastle, 16.08.1947

32. *Aberdeen Press & Journal*, Peat firm send order to Faroe, 16.08.1947

33. Adam Whyte testimony, Report Respecting the Scotch Distilling Duties, 1798, p.96

34. Ian Buxton (ed), *Beer Hunter, Whisky Chaser*, Neil Wilson Publishing, 2009, p.35

35. Charles MacLean, *Whisky Tales*, Little Books Ltd, London, 2006, p.108

Postscript

1. Stuart McLean, *BLACK GOO: Forceful Encounters with Matter in Europe's Muddy Margins*, Cultural Anthropology, 13.10.2011

2. John Buchan, *The Island of Sheep*, Hodder & Stoughton, 1936

3. *Føroyar* translates as "sheep islands" in Faroese

4. E Linklater, Foreword. In: Williamson, K., *The Atlantic Islands: A Study of the Faeroe Life and Scene*, Collins, London, 1948, p.16

5. Archibald Geikie, *Scottish Reminiscences*, James Maclehose & Sons, Glasgow, 1904, p.285

6. Cited in James Murray, *Jim Murray's Complete Book of Whisky*, Carlton, 1997, p.73

7. Petur Jacob Sigvardson, *"Torvið í Føroyum - í søgu og siðsøgu"*, Peat in the Faroe Islands in History and Cultural History, Summary in English, 2006, p.1923-1962

8. John S Owen, *Coal Mining at Brora 1529-1974*, Highland Libraries, 1995, p.30

9. Einar released his first single malt in 2020

Distilleries and Distillers
Featured in the Book

The distilleries and distillers are listed alphabetically; chapter numbers in brackets.

Maps and Illustrations

Maps are © Saraband and/or dmaps.com. Unless otherwise noted above, images are from Saraband's image collection or in the public domain.

INDEX

INDEX

INDEX

INDEX